REVIEWS OF UNITED KINGDOM STATISTICAL SOURCES

Volume XXVI

INTERNATIONAL ASPECTS OF UK ECONOMIC ACTIVITIES

REVIEWS OF UNITED KINGDOM STATISTICAL SOURCES

Editors: W.F. Maunder
and M.C. Fleming

Volume I	*Personal Social Services*, B. P. Davies *Voluntary Organizations in the Personal Social Services Field*, G. J. Murray
Volume II	*General Government Routine Health Statistics*, Michael Alderson *Social Security Statistics*, Frank Whitehead
Volume III	*Housing in Great Britain*, Stuart Farthing *Housing in Northern Ireland*, Michael Fleming
Volume IV	*Leisure*, F. M. M. Lewes and S. R. Parker *Tourism*, L. J. Lickorish
Volume V	*General Sources of Statistics*, G. F. Lock
Volume VI	*Wealth*, A. B. Atkinson and A. J. Harrison *Personal Incomes*, T. Stark
Volume VII	*Road Passenger Transport*, D. L. Munby *Road Goods Transport*, A. H. Watson
Volume VIII	*Land Use*, J. T. Coppock *Town and Country Planning*, L. F. Gebbett
Volume IX	*Health Surveys and Related Studies*, M. Alderson and R. Dowie
Volume X	*Ports and Inland Waterways*, R. E. Baxter *Civil Aviation*, Celia M. Phillips
Volume XI	*Coal*, D. J. Harris *Gas*, H. Nabb *Electricity Industries*, D. Nuttall
Volume XII	*Construction and the Related Professions*, M. C. Fleming
Volume XIII	*Wages and Earnings*, Andrew Dean
Volume XIV	*Rail Transport*, D. H. Aldcroft *Sea Transport*, Derrick Mort
Volume XV	*Crime*, Monica A. Walker
Volume XVI	*Iron and Steel*, David W. Heal *Shipbuilding*, Anthony Slaven
Volume XVII	*Weather*, B. W. Atkinson *Water*, E. C. Penning-Rowsell and D. J. Parker
Volume XVIII	*Posts and Telecommunications*, S. Wall and P. Nicholson
Volume XIX	*Intellectual Property Rights*, D. L. Bosworth
Volume XX	*Religion*, L. Barley, C. Field, B. Kosmin and J. Nielsen
Volume XXI	*Finance*, K. Phylaktis and G. Kaye
Volume XXII	*Printing and Publishing*, W. D. McClelland
Volume XXIII	*Agriculture*, G. H. Peters
Volume XXIV	*Local Government*, J. M. Gillespie
Volume XXV	*Family Planning*, P. F. Selman
Volume XXVI	*International Aspects of UK Economic Activities*, Peter J. Buckley and Robert D. Pearce

Volumes I–XXII are available from Pergamon Press, Headington Hill Hall, Oxford OX3 0BW.
Volumes XXIII–XXVI are available from Chapman & Hall Ltd,
2–6 Boundary Row, London SE1 8HN.

REVIEWS OF UNITED KINGDOM STATISTICAL SOURCES

Edited by W. F. MAUNDER
Professor Emeritus of Economic and Social Statistics,
University of Exeter

and M.C. FLEMING
Professor of Economics,
Loughborough University

Volume XXVI

INTERNATIONAL ASPECTS OF UK ECONOMIC ACTIVITIES

PETER J. BUCKLEY
Professor of Management Economics,
Bradford University

and

ROBERT D. PEARCE
Senior Research Fellow,
Reading University

Published for The Royal Statistical Society and
The Economic and Social Research Council

CHAPMAN & HALL
London · New York · Tokyo · Melbourne · Madras

UK Chapman & Hall, 2-6 Boundary Row, London SE1 8HN

USA Chapman & Hall, 29 West 35th Street, New York NY10001

JAPAN Chapman & Hall Japan, Thomson Publishing Japan, Hirakawacho Nemoto Building, 7F, 1-7-11 Hirakawa-cho, Chiyoda-ku, Tokyo 102

AUSTRALIA Chapman & Hall Australia, Thomas Nelson Australia, 102 Dodds Street, South Melbourne, Victoria 3205

INDIA Chapman & Hall India, R. Seshadri, 32 Second Main Road, CIT East, Madras 600 035

First edition 1991

Softcover reprint of the hardcover 1st edition 1991

Typeset at Oxford University Computing Service

British Library Cataloguing in Publication Data

Buckley, Peter J. (Peter Jennings) *1949–*
International aspects of UK economic activities. –
(Reviews of United Kingdom statistical sources; v. 26).
1. Economics
I. Title II. Pearce, Robert D. (Robert Desmond) *1943–*
III. Royal Statistical Society IV. Economic and Social Research Council V. Series
330

ISBN-13: 978-94-011-6954-7 e-ISBN-13: 978-94-011-6952-3
DOI: 10.1007/978-94-011-6952-3

Library of Congress Cataloging-in-Publication Data

Buckley, Peter J., 1949–
International aspects of UK economic activities /
Peter J. Buckley and Robert D. Pearce. — 1st ed.
p. cm. — (Reviews of United Kingdom statistical sources : v. 26)
Includes index.
ISBN-13: 978-94-011-6954-7
1. Great Britain—Foreign economic relations—Statistics.
I. Pearce, R.D. (Robert D.) II. Royal Statistical Society (Great Britain) III. Economic and Social Research Council (Great Britain)
IV. Title. V. Series.
HF1533.B77 1990
337.41–dc20 90–42888
CIP

VOLUME CONTENTS

FOREWORD

The Sources and Nature of the Statistics of the United Kingdom, produced under the auspices of the Royal Statistical Society and edited by Maurice Kendall, filled a notable gap on the library shelves when it made its appearance in the early post-war years. Through a series of critical reviews by many of the foremost national experts, it constituted a valuable contemporary guide to statisticians working in many fields as well as a bench-mark to which historians of the development of Statistics in this country are likely to return again and again. The Social Science Research Council (now the Economic and Social Research Council) and the Society were both delighted when Professor Maunder came forward with the proposal that a revised version should be produced, indicating as well his willingness to take on the onerous task of editor (a task in which he was assisted from 1985 by Professor Fleming). The two bodies were more than happy to act as co-sponsors of the project and to help in its planning through a joint steering committee. The result, we are confident, will be judged a worthy successor to the previous volumes by the very much larger 'statistics public' that has come into being in the intervening years.

DAVID STAFFORD
Secretary
Economic and Social Research Council

B. W. SILVERMAN
Honorary Secretary
Royal Statistical Society

MEMBERSHIP OF JOINT STEERING COMMITTEE

(December 1987)

Chairman: Miss S. V. Cunliffe

Representing the Royal Statistical Society:

Mr M. C. Fessey

Dr S. Rosenbaum

Mrs E. J. Snell

Representing the Economic and Social Research Council:

Mr I. Maclean

Miss J. Morris

Secretary: Mr D. E. Allen

EDITORIAL INTRODUCTION

The series of *Reviews of United Kingdom Statistical Sources* is meant to serve a dual purpose: to provide an authoritative guide to statistical sources in the UK (both official and unofficial) and a critical appraisal of the nature and limitations of the available data. To maximise its usefulness as a source of reference, each volume in the series follows a standard format which incorporates a number of features designed to meet the varied needs of different users. A detailed guide for users follows this introduction.

A review of statistics relating to international aspects of the economic activities of the United Kingdom was high on the list of priorities of the Steering Committee for the series for some years. At a time when the future prosperity of the country seems to depend increasingly on its success in overseas markets and when the advent of the Single European Market in 1992 is very close, the field it covers is especially important. The scope of the volume has been pitched widely to cover many different aspects of the inter-relationship of the UK with the outside world. The central subjects are, of course, statistics of international trade and the balance of payments, but coverage also extends to related topics such as the holdings of external assets, exchange rates, foreign aid, technical assistance and international flows of labour.

Regrettably, this volume will be one of the last in the series to appear under the joint sponsorship of the Royal Statistical Society and the Economic and Social Research Council (ESRC). The work received the generous financial support of the ESRC (formerly the Social Science Research Council) from the start of the series in 1969 until the end of 1987. During that time work on 25 volumes covering 43 review topics was completed and work on four other reviews had been carried near to completion. The original objective was to try to cover every field of economic and social statistics in the United Kingdom; unfortunately its achievement proved to be a more lengthy task than was originally anticipated and the series is still incomplete. However, the need for such a series and, moreover, a series which is regularly updated remains as strong as ever. More than twenty areas remain uncovered and many of the early volumes now stand in need of extensive revision. Although a valuable *Guide to Official Statistics* is now regularly published under government auspices every two years or so, it does not provide a substitute for the series because, valuable though it is, it is confined very largely to *current* sources of *official* statistics and, perhaps more important, it ventures no appraisals to assist in the use and interpretation of the available data. It is thus a complement to this series rather than a substitute for it. One must, therefore, hope that some further financial sponsorship may yet prove forthcoming which will enable the series to be continued in one form or another.

The editorship of the series was in the hands of Professor W. F. Maunder throughout the whole period to the end of 1987. This volume was completed subsequently and it falls to me, having taken over as editor in order to see the volumes remaining in the pipeline through to publication, to write this Introduction. It remains appropriate, however, that Professor Maunder's name should appear as an editor on this, the twenty-sixth volume, because so much of the work had been accomplished before he stepped down. It is also appropriate perhaps that it should fall to me, as assistant editor during the latter part of the period and, indeed, as an author of two reviews in the series, to pay tribute here to his labours over such a lengthy period—a period not only long but also, at times, difficult. The series owes a great deal to his original foresight and to his perseverance, patience and determination.

It only remains to express gratitude to the members of the Joint Steering Committee of the Royal Statistical Society and the Economic and Social Research Council, who have directed the overall strategy with as admirable a mixture of guidance and forbearance as any editors of such a series could desire. At the same time they bear no responsibility for shortcomings in execution. Especial thanks are due to David Allen, the Secretary of the Committee, who was involved with the project almost as long and almost as closely as anybody. The authors join us in thanking as well all those who gave up their time to attend the seminar which was held to discuss the first draft of the review and which contributed materially to improving the final version.

We are most grateful to Chapman and Hall Limited for their support and in particular to their production department, who put all the pieces together. The subject index entries have been compiled by Mrs Marian Guest who also acted as editorial assistant until October 1989. Linda Lilburne then took over at short notice, very ably, and looked after the final stages. Special thanks are also due to Mr Ray Burnley of the Social Studies Data Processing Unit at the University of Exeter, who again has masterminded our use of the Lasercomp System at Oxford University Computing Service, and to the latter for the use of this facility. We are also very grateful indeed to the Library Association for help in meeting the editorial costs of preparing this volume for publication. Finally, we also wish to record our appreciation of the permission granted us to reproduce certain copyright material by the Controller of Her Majesty's Stationery Office.

Michael Fleming
Loughborough University

GUIDE TO THE SERIES AND HOW TO USE IT

The Scope and Nature of the Series

The purpose of the series is twofold. The primary aim is act as an authoritative work of reference to the sources of statistical material of all kinds, both official and unofficial, in the United Kingdom. The intention here is to enable the user to discover what data are available on the subject in which he or she is interested and from where they may be obtained. The second aim is to provide a critical appraisal of the nature and limitations of the available data so that the user is able to interpret them safely and avoid pitfalls in their use.

Data are regarded as available not only if published in the normal printed format but also if they are likely to be released to a *bona fide* enquirer in any other form, such as duplicated documents, computer print-out or even magnetic tape. On the other hand, no reference is made to material which, even if it is known to exist, is not accessible to the general run of potential users. The distinction, of course, is not clear-cut and mention of a source is not to be regarded as a guarantee that data will be released; in particular cases it may well be a matter for negotiation. The latter caution applies with particular force to the question of obtaining computer print-outs of custom-specified tabulations.

The intention is that the source for each topic should be reviewed in detail, and the brief supplied to authors calls for comprehensive coverage at the level of 'national interest'. This term does not denote any necessary restriction to statistics collected on a national basis (still less, of course, to national aggregates) but it means that sources of a purely local character, without wider interest in either content or methodology, are excluded. Indeed, the mere task of identifying all material of this latter kind is an impossibility. The interpretation of the brief has obviously involved discretion and it is up to the users of these reviews to say what unreasonable gaps become apparent to them. They are cordially invited to do so by communicating with the Editor.

The Format and Content of Reviews

To facilitate the use of the series as a work of reference, a common format is adopted for each review. This involves the incorporation of six standard features:

1. Text
2. Quick Reference List (QRL)
3. QRL Key to Publications
4. Bibliography

5. Subject Index
6. Specimen Forms and Questionnaires

Each of these is described in turn below.

1. The Text. This is designed, in so far as varying subject matter permits, to follow a standard form of arrangement, covering introductory material on the activity covered and its organisation, core material on the available sources and a discussion of desirable improvements.

The introductory material is meant to give a clear background understanding of how data are collected, what is being measured, the stage at which measurements are made, what the reporting units are, the channels through which returns are routed and where they are processed. Coupled with this is a discussion of specific problems of definition and measurement. This is followed by core sections or chapters on available sources. Wherever possible these are arranged according to subject (rather than source). But in practice they may be arranged according to the author's discretion – by origin, by subject subdivision, or by type of data – as there is too much heterogeneity between topics to permit any imposition of complete uniformity on all authors. A final chapter is devoted to a discussion of general shortcomings and desirable improvements. In case a contrary expectation should be aroused, it should be said that authors have not been asked to produce a comprehensive plan for the reform of statistical reporting in the whole of their field. However, a review of existing sources is a natural opportunity to make some suggestions for future policy on the collection and publication of statistics within the scope of their review and authors have been encouraged to take full advantage of it.

2. The Quick Reference List (QRL). This provides a detailed list of all the series and categories of data that are available and, again, is generally arranged according to subject. It also includes cross-references to the sections of the text in which the data are discussed as well as publication references. Each publication shown as a data source is given a serial number and the prefix 'QRL'

3. The QRL Key to Publications. This gives full details of the publications shown as *data sources* in the QRL.

4. The Bibliography. This gives references to works discussing wider aspects of the activity and the statistics including methodology. These publications are identified by a serial number and the prefix 'B'.

5. The Subject Index. This acts as conventional line of inquiry on textual references. But an important feature is that it is a computerised system. For an individual review this introduces the possibility of easy permutation of entries and this is fully utilised in order to facilitate search by giving as many variants as possible. The index is prepared in the editorial office, not by the author.

6. Specimen Forms and Questionnaires. Finally, specimen copies of the more important returns or forms used in data collection are reproduced, as appropriate, as appendices so that it may be seen what tabulations it is possible to make as well as helping to clarify the basis of those actually available.

How to Use the Series

As we have indicated, the standard format adopted for each review in the series is designed expressly to facilitate its use for reference purposes. The features which it incorporates provide three possible 'ways in' for the user. These are:

1. The Contents List
2. The Quick Reference List
3. The Subject Index

For users most interested in discovering quickly whether or not a particular sort of data is available and where it is published, the Quick Reference List should be the most useful feature to consult first. Included within the list are cross-references to discussions and descriptions of the data in the text and references to the publication sources for the data (QRL reference). To facilitate its speedy use, it is arranged by subject as far as possible and the list is itself preceded by its own summary contents list.

The contents list provides, of course, a summary indication of the subject matter of each chapter but also shows the main sub-divisions into which it is divided. It is useful, therefore, for anyone interested in locating material in the book on broad subject areas, rather than more specific statistical sources.

The subject index provides an alternative means of locating information but one which is intermediate between the Quick Reference List and the contents list. It gives references to the subject matter at a more detailed level than the contents list, but it does not duplicate the degree of detail of the entries, relating to particular categories of statistical data, given in the QRL.

44: INTERNATIONAL ASPECTS OF UK ECONOMIC ACTIVITY

PETER J. BUCKLEY
Professor of Managerial Economics
University of Bradford

ROBERT D. PEARCE
Senior Research Fellow
Department of Economics
University of Reading

REFERENCE DATE OF SOURCES REVIEWED

This review is believed to represent the position, broadly speaking, as it obtained in August 1987, but references have been included to major changes introduced in January 1988. An addendum on more recent developments up to January 1989 is included at the end of the volume immediately preceding the Appendices.

CONTENTS OF REVIEW 44

ACKNOWLEDGEMENTS

Our primary debts are to Professor W. F. Maunder and Professor Michael Fleming, editors of this series and to Marian Guest (Editorial Assistant) who have shepherded us through the minefield to completion.

We have been fortunate in the range of people who have helped us, guided us and commented on earlier drafts. These people include: Mr A. H. Cowley of H M Customs and Excise Statistical Office, Mr B. J. Buckingham of the Central Statistical Office, Mr W. Hall, Mr R. J. Walshaw and Mr A. Hewer, of the Department of Trade and Industry, Dr David Mayes of NEDO, Mr S. J. Peachey of the Bank of England, Dr Eric Pentecost of Loughborough University (formerly of the Bank of England), Mr Tony White, Professor V. N. Balasubramanyam of the University of Lancaster, Dr Peter Enderwick of Queen's University, Belfast, Professor Alan Winters of University College of North Wales, Bangor, Dr Kirsty Hughes of the University of Manchester and Mrs Jocelyn M Slee of Special Chemical Statistics Limited. Mr Neil Hunter and the staff of the library of the University of Bradford Management Centre have been unfailingly helpful.

Thanks are also due to the Steering Committee, Mr Martin Fessey, who chaired an excellent seminar discussion on an earlier draft, Mr Ian MacLean, Dr Sidney Rosenbaum and Mr David Allen of ESRC.

We would also like to thank Mrs Marian Taylor for research assistance and Mrs Christine Barkby, Miss Lyndsey Collins, Mrs Sylvia Ashdown, Mrs Jill Turner and Mrs Pat Elgar for typing and retyping the myriad drafts.

All of the above are absolved from responsibility for any remaining mistakes. Without their help, the mistakes would have been far more plentiful.

P.J.Buckley and R.D.Pearce

LIST OF ABBREVIATIONS

ABI	Association of British Insurers
ATP	Aid and Trade Provision
BBC	British Broadcasting Corporation
BEC	(United Nations Classification) Broad Economic Category
BESS	British Expatriates Supplementation Scheme
BIS	Bank for International Settlements
BSO	Business Statistics Office
BTN	Brussels Tariff Nomenclature
CCCN	Customs Cooperation Council Nomenclature
CCT	Common Customs Tariff
CDC	Commonwealth Development Corporation
CF	Consolidated Fund
CGBR	Central Government Borrowing Requirement
cif	Cost, Insurance, Freight
CMIs	Central Monetary Institutions
CSO	Central Statistical Office
DAC	Development Assistance Committee
DTI	Department of Trade and Industry
EC	European Community
ECGD	Export Credit Guarantee Department
ECU	European Currency Unit
EEA	Exchange Equalisation Account
EFTA	European Free Trade Association
EMS	European Monetary System
ERI	(Sterling) Exchange Rate Index
FAO	Food and Agriculture Organisation
fob	Free on Board
GATT	General Agreement on Tariffs and Trade
GDP	Gross Domestic Product
GNP	Gross National Product
HGV	Heavy Goods Vehicle
HMG	Her Majesty's Government
HS	Harmonised System
IBA	Independent Broadcasting Authority
IBRD	International Bank for Regional Development
IDA	International Development Association
IFAD	Internatonal Fund for Agricultural Development

IFC	International Financing Corporation
IMF	International Monetary Fund
IPD	Interest, Profits and Dividends
IPS	International Passenger Survey
IRM	Institute for Research and Information on Multinationals
LIC	Low Income Countries
LLDC	Least-Developed Countries
LMDA	London Discount Market Association
LMIC	Lower Middle Income Countries
MERM	Multilateral Exchange Rate Model
MLH	Minimum List Heading
MMC	Main Manufacturing Countries
NEDO	National Economic Development Office
NEI	Not elsewhere identified
NIMEXE	Nomenclature of Goods for External Trade Statistics of the European Community and Statistics of Trade between Member States
NLF	National Loans Fund
ODA	Official Development Assistance
OECD	Organisation for Economic Cooperation and Deveolopment
OOF	Official Other Flows
OSAS	Overseas Service Aid Scheme
OTS	Overseas Trade Statistics
PE	Public Expenditure
SDR	Special Drawing Rights
SIC	Standard Industrial Classification
SITC	Standard Industrial Trade Classification
SNAPS	Ships, North Sea Installations, Aircraft, Precious Stones, Silver
TARIC	Integrated Customs Tariff of the European Community
TC	Technical Cooperation
TOF	Total Official Flows
TTCN	Tariff Trade Code Number
UMIC	Upper Middle Income Countries
UN	United Nations
UNDP	United Nations Development Programme
VAT	Value Added Tax
VFR	Visits to Friends and Relatives

OUTLINE

0.1 Introduction

0.1.1 *Overview*

This volume surveys international aspects of UK economic activities. Its concern is to help in the process of transforming data into information. It has a central focus on official statistics but takes in private, non-governmental work where appropriate.

0.1.2 *Structure of the Volume*

Chapter 1 outlines the UK balance of payments. It attempts to cover the principal components of the accounts and to give estimates of the reliability of the statistics as far as this is possible.

Chapter 2 examines UK visible trade, its classification (and changes in that classification), composition, direction and valuation. It also covers measures of trade performance such as import penetration and export sales ratios.

Chapter 3 covers UK invisible trade: trade in services, earnings on UK external assets and liabilities (interest, profit and dividends) and transfers. There are many problems in accounting for invisible items and the chapter attempts to assess these where appropriate.

Chapter 4 is a transitional chapter linking the balance of payments to later chapters on individual elements of capital movements. It examines transactions in UK external assets and liabilities and levels of UK external assets and liabilities.

Chapter 5 examines in detail outward and inward direct investment, both annual transactions and book values. The primary problem in the material covered in this chapter and chapter 6 is constructing consistent time series and much of the concern of the chapter is to establish the location of information and its compilation into consistent series.

Chapter 6 has a similar orientation in its coverage of royalties and miscellaneous overseas transactions. As well as royalties, the chapter covers film and television transactions and direct exports included in the DTI annual survey of overseas transactions.

Chapter 7 examines information on the international transactions of the UK monetary and banking sector.

Chapter 8 covers sterling exchange rates, concentrating on sources which provide reasonably long time series.

UK foreign aid is covered in Chapter 9 together with related technical assistance. The main components are analysed together with gross expenditure, technical cooperation expenditure, debt relief, Commonwealth Development Corporation transactions, technical cooperation manpower data and detailed information on the comparative performance of UK foreign aid.

Chapter 10 examines international flows of labour of the UK, including international migration.

Chapter 11 considers desirable improvements and future developments.

0.1.3 *Coverage*

For the purposes of trade statistics, the 'United Kingdom' (UK) is defined as Great Britain, Northern Ireland, the Isle of Man, Channel Islands and the UK part of Continental Shelf.

0.1.4 *International Comparability*

The balance of payments is the record of an economy's international economic transactions, i.e. the record of the goods and services that an economy has received from and provided to the rest of the world and of the changes in the economy's claims on, and liabilities to, the rest of the world [B.26].

The balance of payments statement can be put to a variety of uses, perhaps the primary one being an appraisal of the effects of international transactions on the domestic economy. The IMF *Balance of Payments Manual* [B.26] shows that analytic presentations of the balance of payments can be constructed by classifying its standard components into two or more categories according to the relevant attributes chosen for examining some particular aspect of a country's international economic relationships. The most usual of these are: (1) balance on goods, services, income and unrequited transfers commonly referred to as the current balance or current account; (2) the balance of payments. The latter, based on the principles of double entry book-keeping, conceals many inter-relationships between separate entries and must be interpreted with caution by analysts and policy-makers. There are also strong causal relations between the overseas sector and domestic economic variables. In no sense are entries in the accounts separate or separable from events in the British economy or in the rest of the world. This is particularly true where multinational (or transnational) firms span national borders. The choice of technique of servicing foreign markets and the input-purchasing or sourcing policies of these firms affect the level of economic activity in individual nation states. Flows of international finance link the capital markets of individual nations and the transactions of Government increasingly have international implications. Other usual presentations are: (3) Balance financed by reserve assets and selected liabilities, often termed an 'overall' balance and (4) balances on goods, services, income, unrequited transfers and long-term capital movements. The final balance is often useful where the current balance is too narrow and this so-called 'basic' balance picks up longer-run trends. The overall balance has, of course, important implications for the official authorities and their financing policies.

Consequently, the balance of payments is made up of standard components which are the basic building blocks for various types of analytical presentation.

In principle, the world current account must sum to zero. In practice it does not do so and the discrepancy is growing [B.17]. Basically, this problem arises from changes in the nature and speed of international financial transactions and the use of new financial instruments. The closer integration of international financial markets and liberalisation has increased flows and there has also been an increasing concealment by investors. The use of indirect channels such as multinational banks, offshore funds and financial institutions based in tax havens complicates the process of recording and attributing capital flows in individual country accounts. It is service transactions which pose the main difficulties rather than goods flows (see Chapter 3 for a review of the UK situation). Within this, it is the failure of investment income accounts to capture cross-border assets accurately which is the key problem. The growth of multinational corporations and their increasingly inter-related and complex organisations make geographical attribution of direct investment income problematic. Re-invested income of multinational firms is difficult to record accurately particularly as many host countries do not collect or estimate the requisite information. International portfolio income also poses problems — inflows of capital reported by countries receiving funds have been much greater than those reported in the accounts of the country of origin. Debtor countries, recognising the full extent of these liabilities, record reasonably fully the interest payable to non-resident owners whereas the source countries have little or no record of the accumulating capital and do not record a comparable amount of income credit. This is the largest element in the global current account discrepancy. Its import should be borne in mind when examining the relevant sections of this volume. Shipping and transportation also contribute to the discrepancy, but this element is relatively static, not growing. Unrequited transfers record more debits than credits globally also. In response to these problems the IMF is to update its *Balance of Payments Manual* in order to catch up with these developments. This will undoubtedly affect future UK presentations.

Foreign Investment Statistics. Coverage of foreign investment globally on a consistent basis is a large problem as the preceding section has shown. In particular, coverage of foreign direct investment differs by country in terms of coverage, scope, definitions, methodology, disaggregation and periodicity. In view of this it us essential to note a general point: UK figures on outward foreign direct investment in country X will not correspond with any degree of exactitude with country X's figures on inward foreign direct investment. Part of this arises from definitional differences. As the defining characteristic of direct investment is foreign control, this is translated into different proportions of foreign equity ownership by different countries. The OECD has suggested a common proportion of equity capital (ordinary voting shares) in foreign hands should be 10 percent to define the existence of a foreign direct investment relationship. However, this should be treated 'flexibly' (OECD: *Detailed Benchmark Definition of Foreign Direct Investment* [B.34]).

Foreign Direct Investment is generally, as in the UK, measured at book value, rather than replacement or market values. The latter are unlikely to be available in the near future.

Summary. Efforts to harmonise statistics of international aspects of economic activity across countries are increasing. Within the European Communities the advent of the 'Single European Act' is having a major impact on the collecting and presentation of data and this will increase. Efforts in the international institutions, notably the IMF and OECD, are also leading to the alignment of definitions, coverage, methodology and presentation. However, given the scale of the problems and the immensity of the task, together with the accelerating pace of change, it is likely to be some time before harmonisation overcomes national idiosyncracy.

CHAPTER 1

THE UNITED KINGDOM BALANCE OF PAYMENTS

1.1 General

The balance of payments is the statistical record of all economic transactions which have taken place during a given period of time between the residents of one country and the rest of the world. The transactions may represent flows of resources provided by or to the country's residents (goods and services exported and imported and the use of investments), changes in the country's foreign assets or liabilities, and transfer payments.

In principle, transactions are recorded when the ownership of goods or assets changes and when the services are rendered. In practice this is not always possible.

1.1.1 *Structure*

In the UK balance of payments accounts — the 'Pink Book' — [QRL.61] transactions are classified into two main groups as follows:

1. *Current account transactions* which cover exports and imports of goods and services, investment income and transfers.
2. *Transactions in UK external assets and liabilities* which cover inward and outward investment; external transactions by banks in the United Kingdom; borrowing and lending overseas by other UK residents; drawing on and accruals to the official reserves, and other capital transactions.

The *current balance* shows whether the United Kingdom has had a surplus of income over expenditure, and, taken with capital transfers, it shows whether the United Kingdom has added to or consumed its net assets in any period.

Economic transactions giving rise to demand for foreign (non-sterling) exchange result in debit entries and transactions increasing the supply of foreign exchange result in credit entries. As the 'Pink Book' puts it 'under the conventions of the balance of payments accounts credit (or positive) entries are recorded for resources provided: these may be exports of goods or services or the use abroad of UK investments. Correspondingly, debit (or negative) entries are recorded for resources provided to UK residents' ([QRL.61] 1987 p. 3).

1.1.2 *Residence*

The UK balance of payments records transactions between non-residents and residents of the United Kingdom, including the Isle of Man, Channel Islands and that part of the Continental Shelf which is defined by international convention as belonging to the United Kingdom. The term 'resident' covers both individuals living

Table 1.1

The Current Account and Transactions in External Assets and Liabilities and the Balance of Payments: Sign Conventions

	Domestic Residents	*Foreign Residents*
Domestic goods and services	No balance of payments effect	Purchase (+)
Foreign goods and services	Purchase (−)	No balance of payments effect
Transfer payments to residents	No balance of payments effect	Debit (−)
Transfer payments from residents	Credit (+)	No blance of payments effect
Income on foreign assets	Credit (+)	No balance of payments effect
Income on domestic assets	No balance of payments effect	Debit (−)
Foreign asset transactions	Purchase (−) Sale (+)	No balance of payments effect
Domestic asset transactions	No balance of payments effect	Purchase (+) Sale (−)
(See also Pentecost [B.19])		

permanently in the United Kingdom and corporate bodies located in the United Kingdom, but not their overseas branches and subsidiaries.

Individuals, other than government officials, members of the Armed Forces and students, who enter or leave the country with the intention of staying here or abroad for a year or more, are regarded as migrants changing their residence (when Exchange Control regulations were in force, three years rather than one was used as the criterion for residence). Government agencies operating abroad, such as Embassies or military units, are regarded as residents of their home countries.

Overseas branches of companies are treated as residents of the country in which they are located and subsidiaries are treated as residents of the country in which they are registered. Agencies of business enterprises are also regarded as residents of the countries in which they operate. Transactions through agents are, in general, treated as though made directly with the overseas principal, but commissions etc. received by agents for their services are regarded as accruing to the agent's country of residence. For a detailed description of the criteria adopted for defining residence see [B.43].

1.1.3 *Accounting Principles*

The principles of balance of payments data collection and presentation are those of double-entry book-keeping. Consequently, every transaction must enter one account with a credit (left hand side) and another with a debit (right hand side). Instead of the practice of arranging in credit and debit columns which sum to the same total, balance of payments items are listed in a single column, but given different signs so that they sum to zero, thus accommodating items which are generally recorded only on a net basis. Also, analytical balances of groups of transactions may be displayed more readily in this form of presentation. Typically (but not inevitably) one half of the transaction will relate to resources and the other half to the financing item.

Since the two entries made in respect of each transaction are generally derived from separate sources and the methods of estimation are neither complete nor precisely accurate, the two entries may not match each other precisely or may fall within different recording periods. Furthermore, with timing variations it is possible that entries which correspond in foreign currency terms, are converted to sterling at different rates of exchange. In order to bring the total of all entries to zero, a *balancing item* is included to reflect the sum of all these errors and omissions.

There is an unfortunate and unwarranted normative connotation given to payments balances. The implication or inference that a positive balance is 'good' and a negative balance 'bad' is often drawn, particularly in the reporting of quarterly current account figures. A current account surplus means that the UK is sending abroad a net amount of goods and services overseas. These goods and services are thus not available for domestic consumption and investment. In a current account surplus, the UK is receiving a claim on assets abroad in return for these net exports.

1.1.4 *Interdependencies*

Interdependencies should also be noted. There are interdependencies between different elements of the balance of payments accounts. Picking on one element of the accounts and inferring causality is extremely dangerous. The general equilibrium nature of the accounts means that immediate inferences can be fallacious. For instance, the balance of payments may have deteriorated by $100 million, whilst long-term deterioration of £100 million also occurred in the long-run capital account. It might thus be argued that the deficit was *caused* by the outflow of long-term capital. This ignores interdependencies between different classes of transaction and with economic aggregates in the home and world economies. The outflow of capital may have been a direct investment undertaken to purchase a foreign company which, in turn, purchased £200 million of equipment from the UK, thus increasing UK activity and exports and the current account balance.

The second set of interdependencies is between the UK and the rest of the world. Theoretically, the world's balance of payments should sum to zero. However, the global current account has shown a growing discrepancy, from a rough balance in 1970, to an imbalance of £20 million in 1978, £109 billion in 1982 although it declined to £56 billion in 1986 (see [B.17], [B.46]). The IMF Executive Directors set up a Working Party to examine the reasons for this which reported in 1987 [B.37]. Essentially, the discrepancies result from major changes in the nature and speed of international financial transactions, notably new financial instruments. This, combined with the liberalisation of the international financial system and closer integration of financial markets and rapid changes in the technology for effecting transactions have led to statistical gaps in the income sector of the global balance of payments even after reconciling on to a uniform (IMF) basis. The emergence of these growing 'cross-border assets' — capital held in other countries — is often recognised by debtor nations but not by creditor nations. Direct investment income, reinvested earnings from direct investment and interest, dividends, profits are difficult to measure and estimation techniques are often arbitrary. The IMF working party [B.37] showed that the largest and fastest growing discrepancy, however, was portfolio investment income (including suppliers' credit). This accounted for 43 per cent of the gap in reporting of income from services in 1984. A substantial part of world freight earnings and ship expenditures were not covered in the global accounts because of incomplete coverage and specific cases of incorrect reporting. The discrepancy in the shipping and transportation accounts is large, but is relatively steady and so is not likely to lead to errors in current trend analysis. Unrequited official transfers also were often overlooked by some countries whilst private transfers were assymetrically reported by creditors and debtors. The opening up of new offshore banking and financial centres with associated new financial instruments (e.g. swaps and options) tended to obscure final ownership of assets. Changes in banking practice have tended to move these instruments off banks' balance sheets and non-bank asset holders are more difficult to reach so full and accurate figures are difficult to compile. Intermediaries and final owners often have an incentive to conceal their activities.

The IMF working party has suggested corrections by closer examination of accounts and made suggestions for methodological changes [B.37]. It can be expected that national authorities, including the UK, will begin to improve the collection and reporting of data in these difficult and dynamic areas. It is necessary to add that all figures in the area of international financial instruments reported elsewhere in the areas covered by this book should be examined and treated with the utmost caution.

Finally, judgements on trade balances with one trading partner should not be made in isolation from the knock-on effects of altering the balance. In an increasingly interdependent world, bilateral balances have declining significance.

1.2 Statistics: The Balance of Payments Basis

Detailed statistics of the international *trade* element in the balance of payments are published in *Overseas Trade Statistics* [QRL.48], [QRL.49] which are discussed in chapter 2. Certain adjustments can be made to *Overseas Trade Statistics* [QRL.48], [QRL.49] to bring visible trade statistics to a balance of payments basis. These adjustments relate to valuation, coverage and timing.

The principal *valuation* adjustment deducts the freight and insurance elements, which are payments for services, from the c.i.f. (cost, insurance, freight) valuation declared for imports in order to arrive at the required f.o.b. valuation for the balance of payments basis. Trade is, in general, recorded at market values and adjustments are made for certain specific types of transaction known to reflect other valuations. These include certain trade between related companies and, until 1977, trade affected by the EC Common Agricultural Policy.

Some *coverage* adjustments add goods which are not covered by declarations to Customs e.g. second-hand ships and aircraft, new ships delivered abroad and imports directly to North Sea installations. A further coverage adjustment is for those goods in the *Overseas Trade Statistics* [QRL.48], [QRL.49] which do not change ownership and which are therefore removed for balance of payments purposes. Finally, there are goods exported for the benefit of UK forces overseas: these personnel are regarded as UK residents and so these exports are omitted from the balance of payments accounts.

The *timing* of *Overseas Trade Statistics* [QRL.48], [QRL.49] reflects the presentation of Customs documents. This results in statistics of imports for a given calendar month corresponding closely to movements through the ports that month. For exports, the figures relate on average to movements in the monthly period ending about the middle of the given calendar month (see 2.7 below). This timing of recording is believed to be reasonably close to change of ownership and no timing adjustments are made. Goods entering or leaving the UK purely in transit are excluded.

Further details of adjustments from overseas trade statistics to the balance of payments basis are to be found in *UK Balance of Payments* the 'Pink Book' [QRL.61] 1987 edition pages 12-13.

The 'Pink Book' [QRL.61] gives the Summary Balance of Payments on an annual basis covering current account and capital account. Current account entries are: visible balance, invisible balances: services, IPD (interest, profits, dividends), transfers, total and the current balance. Capital transfers have not been entered since 1974. Transactions in UK assets and liabilities are as follows: UK external assets, UK external liabilities and net transactions. The EEA (Exchange Equalisation Account) loss on forward commitments has not been entered since 1968, gold subscriptions to IMF since 1970 and allocation of SDRs since 1981. The balancing item is also given.

1.2.1 *The Balancing Item*

The balancing item in the balance of payments is included to reflect the sum of errors and omissions to bring the total of all entries to zero. It includes persistent elements, where certain types of transaction are not adequately covered in the accounts and erratic fluctuations which are likely to be, in part, timing errors in the recording of transactions and their corresponding payments.

The persistent elements in the balancing item will be affected by changes in the coverage of certain series e.g. deposits with banks overseas have been identified only since the end of 1975 and prior to that would remain in the balancing item. From the end of 1982, a substantial part of trade credit between unrelated companies is no longer identified in the accounts and consequently becomes part of the balancing item. Additionally short-term 'leads and lags' in the timing of payments and small changes in terms of settlement could have a large effect as very large sums are involved in the gross balance of payments. Discrepancies in foreign currency amounts arising from exchange rate changes are also important, particularly when the forward exchange market is used. The balancing item was exceptionally large in 1986 following deregulation of London financial markets ('Big Bang') and concomitant large capital flows. However, much non-bank private sector borrowing from, and investment in, foreign and Euro short-term paper markets is not captured in measured capital flows and may well be contributing to the increasingly large balancing item (see section 1.1.4 and [B.17], [B.46], [B.37]).

1.3 Reliability of Balance of Payments Data

1.3.1 *General*

The balance of payments estimates are compiled from a large number of different sources and the degree of accuracy varies considerably between items. Some of the errors and omissions may be persistent and tend to be in the same direction in all years. For this reason, the absolute error in the change from year to year is likely to be less than the error in any particular year. Errors are likely, to some extent, to offset each other in any particular year, and the proportionate error in major aggregates is likely to be less than the weighted average of the proportionate errors in the components. However, where a balance is drawn between two aggregates and the balance is small in relation to the aggregates, such as the current balance, the proportionate error attached to the balance is liable to be very substantial. The 'Pink Book' [QRL.61] no longer attempts the quantitative grading given in the 1976 issue although a general discussion on reliability is given and the relative gradings within the current account shown in the 1976 edition still have some validity.

The likely reliability of particular components can be judged in what is said in the descriptive notes to each chapter in the 'Pink Book' [QRL.61]; for instance, if an estimate is based on a survey carried out several years ago or is projected from exchange control data, the current figure may be seriously wide of the mark. Fortunately, such instances form a relatively small part of the total account. In relatively few cases are the sample sizes in particular surveys a sufficient guide to the accuracy of the estimates, as reporting errors and other inaccuracies may be equally important. However, for the travel figures (see section 3.2.5), the size of the International Passenger Survey is in large part determined by a wish to limit the sampling error of the expenditure estimates to around 2 per cent (see [QRL.11] 1985 p. 1 and sections 1.3.3 and 3.7 below).

Errors in measuring capital flows also tend to be greater than those in measuring current flows. The scope for error in measuring capital flows has been particularly great in the recent period when there has been a proliferation of new financial instruments and substantial flow of funds to and from the United Kingdom, in part associated with the deregulation of the London Stock Exchange ('Big Bang').

1.3.2 *Reliability: Visible Trade*

A formula for valuing export and import merchandise is laid down on the customs entry forms from which the *Overseas Trade Statistics* are compiled [QRL.48], [QRL.49]. A number of adjustments are made for both valuation and coverage, to bring the *Trade Statistics* figures on to a balance of payments basis (see section 1.2). Some of these adjustments are based on comprehensive returns and others on the results of sample enquiries; any errors in them are thought to be small. For the most important adjustment — the estimate of total freight on dry-cargo imports, which is deducted to bring the value of imports on to an f.o.b (free-on-board) basis — any error will be offset by an equal and opposite error in the invisibles account, since the same estimate is used as the starting point for the calculation of transport freight debits. It is possible that there is some inconsistency between the value of commissions assumed to be included in the figures of exports and imports in the visible trade account and the estimate for credits and debits on commissions included in 'other services' in the invisible account.

In other respects there may be rather less accuracy in the valuation of exports than in the valuation of imports. A particular problem is the case of a series of consignments forming part of a large single contract. In relation to the large gross value involved, the margin of error in the estimates for visible trade is probably very small, although it is still of significance in relation to the difference between the two estimates.

The CSO considers it inappropriate to make an allowance in the visible trade account for changes in the timing of payments for exports and imports. Conceptually, such changes, which may arise through the advancing or retarding of payments through shifts in the timing pattern as a result of changes in the commodity make up of exports and imports, or through shifts to imports on consignment, are regarded as equivalent to short-term capital movements. Such changes in timing of payments are therefore treated in the same way as the giving and taking of normal trade credit. To a large extent trade credit is recorded in the balance of payments accounts, but in so far as changes in the timing of payments are not so recorded, they may contribute significantly to fluctuations in the balancing item, which reflects the total of errors and omissions.

1.3.3 *Reliability: Invisible Trade*

Figures of *government transactions* are based on records of government departments and can be considered to be good; any errors are generally limited to minor timing discrepancies.

On *sea transport* the figures relating to UK shipping operations are based on the annual and quadrennial enquiries carried out by the General Council of British Shipping and the margin of error is unlikely to be large. There may be appreciable errors, however, in the provisional estimates produced before the full results of these inquiries become available.

The figures on freight payments to overseas operators are based on the visible trade adjustment (equal and opposite errors in the estimates of the f.o.b. values of imports). Figures on overseas operators' expenses within the UK are based on data collected several years in advance of publication in the 'Pink Book' [QRL.61] and are subject to some uncertainty.

The estimates on *civil aviation* are considered to be reliable: they are based on regular returns from airlines and airport authorities and on the International Passenger Survey.

Estimates of *travel* are based on the International Passenger Survey (and have been since 1962): a system of interviews with a sample of passengers leaving or arriving at seaports or airports (see section 3.7). It is subject to sampling errors and deficiencies in reporting but the estimates are considered to be satisfactory.

In value terms about three-quarters of the financial and other services estimates are based on direct inquiries of industry and commerce (up from half in 1976) but the estimates for some items are derived from various related indicators and are thus only approximate. The totals for both credits and debits under this heading must be regarded as subject to 'an appreciable margin of error' ([QRL.61], 1987 p. 4).

The *interest, profit and dividends* (IPD) arising from direct investment by companies are thought, by the CSO, to be reasonably satisfactorily measured both on the credit and debit side, although the provisional estimates for the most recent year are less reliable. Estimates of interest and dividends on portfolio investment, however, are much less reliable than those on direct investment. The estimates for certain other elements of IPD vary in quality; this includes oil earnings and other elements of investment income, although measurement of lending and borrowing of overseas currencies has been improved by direct measurement. Other figures are not directly measured.

Estimates of *private transfers* are subject to much uncertainty. For credits, about half the total is based on information from particular countries on their payments to the United Kingdom. The rest is based on various assumptions. For debits, about a third of the total is derived directly from surveys; most of the rest is based on Exchange Control data, trended forward from 1979 using data from the Family Expenditure Survey. It should therefore be treated with extreme caution.

1.3.4 *Reliability: Current Account*

The CSO no longer attempt quantitative gradings as were given up to the 1976 'Pink Book' [QRL.61] page 75. These 1976 estimates are reproduced as Table 1.2

The 1976 'Pink Book' notes (p. 75) that it is 'impracticable to calculate statistical margins of error of the kind that are derived from random samples for any of the aggregates or for most of their components'. The table reproduced as 1.2 was drawn from knowledge of the data to give 'rough and mainly subjective judgements of the range of reasonable doubt attaching to the estimate'. These judgements were presented, in the opinion of the estimators, with the view that there was a 90 per cent chance that the true value of the figures referred to lay within the limits set by the grading. Improvements in the data and estimates since 1976 have been made in the financial and other services areas, but, for the cautious, Table 1.2 represents a reasonable guide to reliability of current account items. In examining time series, it should also be noted that earlier figures might be subject to wider margins of error.

It should be noted that a wider margin of error attaches to the quarterly estimates for most components of invisibles. A table, which is reproduced as Appendix 1, shows the basis of initial quarterly estimates. It is relevant to note the extent to which there is reported data for the current quarter rather than projections from earlier quarterly, annual or other data.

The frequency and scale of revisions in the past is also of interest, although it is not necessarily an exact guide to the current situation as steps are taken to remedy inaccuracies. Table 1.3 shows, as an example, the extent to which the invisibles balance is revised between successive 'Pink Books' [QRL.61].

Table 1.2 Reliability of Annual Estimates of Current Account Credit and Debit Items (from [QRL.61]) (1976)

Category	Grading
Exports and Imports of Goods	A
Invisibles	
Government	A
Sea Transport	B
Civil Aviation	B
Travel	B
Other Services	C
Total Services	B
Total Goods and Services	A
Transfers	
Government	A
Private	C
Total Transfers	C
Interest, Profit and Dividends	B
Total Current Account	B
Current Balance	C

Key: A = less than 3% error margin; B = 3% to 10% error margin;
C = more than 10% error margin

1.3.5 *Reliability: Transactions in External Assets and Liabilities*

Flows of *direct investment* by companies are thought to be adequately measured, but a substantial margin of error attaches to the estimates for the most recent year, which are based on quarterly data because the annual inquiry results, on which they are eventually based, do not become available until about eighteen months after the end of the year. Interim estimates are based on less comprehensive quarterly inquiries.

Following the abolition of exchange control, an inquiry into *portfolio investment* transactions effected through banks and UK security dealers was introduced in mid-1980 by the Bank of England. The inquiry is being supplemented by data on transactions by financial institutions which are available from other statistical inquiries. The estimates are subject to 'an appreciable margin of error' particularly in the most recent period in respect of overseas investment in UK securities.

The figures of transactions in UK banks' assets and liabilities are taken from the regular returns made by banks to the Bank of England. However, errors may arise through failure to recognise non-resident status, especially where nominee accounts are employed. The effects of exchange rate changes have to be allowed for when estimating the sterling value of changes in external assets and liabilities in foreign currencies.

Borrowing from overseas banks by the UK non-bank private sector was originally derived from exchange control sources but since their abolition it is now being covered by data from the Bank for International Settlements (BIS). Here, there may be some element of duplication with existing data such as trade credit. There may also be some inconsistencies of measurement since the BIS figures are collected from individual reporting countries. Again, the effects of exchange rate changes have to be allowed for in translating changes in foreign currency liabilities to sterling. The CSO regards the data as 'being the best available at present', which means that because of potential double counting and omissions, it should be treated with great caution. Similar qualifications apply to *deposits by UK residents other than banks* with banks abroad. BIS data has been used here since 1976, although duplication problems are less severe.

The discontinuation of a Department of Trade and Industry inquiry at the end of 1982 meant that changes in *trade credit* between unrelated companies, always subject to a substantial margin of error, deteriorated further in its accuracy.

In contrast 'fairly complete' records of *general government* and most *public sector capital transactions* are available.

The figures of *other capital transactions* are based on a variety of sources and a number of gaps remain in the estimates for the non-bank private sector in particular.

1.3.6 *Reliability of Balance of Payments Data: Summary*

This section has shown that the reliability of data on individual items in the balance of payments varies enormously. Changes in accuracy have not improved uniformly and in some cases have deteriorated in quality. Because of the wide variety of sources and methods used, there are difficulties in obtaining complete consistency of definition and treatment between one item and another and some overlapping or shortfall results.

The abandonment of quantitative estimates of the reliability of individual items (as Table 1.2) is a retrograde step and its reinstatement would be a boon to users.

It is essential that users of balance of payments data scrutinise the notes to tables in the 'Pink Book' [QRL.61], particularly when compiling time series. Changes in content and revisions should be noted with care.

Table 1.3 Revisions to the Invisible Balance (£million)

Changes between	1986	1985	1984	1983	1982
March and June	+979	+811	+573	+875	+133
June and (first) Pink Book	-650	-118	+157	+208	+1456
First and second Pink Books		-616	+560	+371	-487
Second and third Pink Books			+262	-34	-490
Third and fourth Pink Books				+204	-726
Fourth and fifth Pink Books					+98

The changes shown in this table indicate the extent to which the initial estimate for a particular year is subsequently revised. Although there has been a tendency for the balance to be revised upwards between March and June of the following year, thereafter the changes have been in both directions. The classic case is 1982 where the balance is now virtually the same as it was initially in March 1983 after having been revised upwards by £1.6 billion and then progressively downwards by the same amount! In contrast, 1984 has been revised progressively upwards and is currently £1.6 billion above the initial estimate.

1.4 The Current Account

The publication of monthly, quarterly and annual figures for balance of payments statistics represents a convergent process of estimation. However, there is often too much attention paid by commentators to short-term fluctuations. There are far too many random events which impinge upon the magnitudes measured in the accounts to make any single month's or quarter's figures significant. The revision of quarterly figures for up to 5 quarters after initial publication can alter the figures significantly. It usually takes a number of quarters, even years, for significant trends to be discerned.

Quarterly estimates of the UK balance of payments are published by press notice each quarter [QRL.62] and detailed tables are included with the press notice text in *Economic Trends* (March, June, September and December) [QRL.30]. Those published in September are consistent with the annual figures in the 'Pink Book' [QRL.61]. These latest quarterly estimates are also reproduced in summary form in the *Monthly Digest of Statistics* [QRL.42] and, for the summary account and for the capital account in detail in *Financial Statistics* [QRL.33]. Monthly estimates of visible trade and projected invisible balance (expressed as a monthly rate) seasonally adjusted are published in a DTI/CSO *Press Notice* each month [QRL.25] and subsequently in *British Business* [QRL.7].

The current account of the UK balance of payments is given on a monthly basis in a DTI and CSO press notice [QRL.25]. Some information on invisibles is also given. The press notice gives details of the most recent period, more comprehensive coverage of earlier periods being contained in [QRL.43]. The publication of the DTI/CSO press notice [QRL.25] includes a projection of the net balance on invisibles to give a broad indication of the current account. Although the invisibles balance is expressed as a monthly figure, there is virtually no monthly information on invisibles. The figures are no more than quarterly estimates or projections expressed at a monthly rate. These projections are reassessed in the light of accumulating evidence. The initial projection is strongly influenced by trends in the recent past. The single published figure is prepared in terms of the main components (not published). Explicit assumptions about interest rates and exchange rates are made for the rest of the quarter (at the time the projection is made, around mid quarter point). If there is evidence on the timing of particularly large transactions (e.g. payments to or from the European Community or in respect of overseas aid), this is also taken into account. The uncertainty about the projection of the invisible balance results from the variability of the balance of payments estimates — it is the difference between two large and variable estimates of debits and credits. Relatively small differences between rates of changes in credits and debits can lead to a huge change in the net balance. The aggregate balance is published only to the nearest £100 million to stress its approximate character. Occasionally, when a large element is included, some additional details are given to explain why the latest estimate is substantially out of line with the immediate past. For details of projection methods for the monthly invisible balance see [B.35]. The monthly invisible balance lacks the precision of the visible trade figures, as invisible transactions do not pass any single point such as customs where they can be consistently measured. The variety of sources and the variability in time periods of measurement (very occasionally monthly, usually quarterly and more comprehensively and exhaustively, annually) means that the data must be treated with extreme caution. The uncertainty remains even though the projection is reviewed systematically in the light of accumulated information.

The value of exports and imports is given based on commodity classes of SITC (Rev 2) section, division or group and UN Classification by Broad Economic Categories (BEC). Volume indices of export and import volume are given on the same basis (1980 = 100) as are unit value indices and the value of export and imports by areas of the world.

The 'Pink Book' [QRL.61] gives a summary of the current account on an annual basis: credits, debits and balances. This table is repeated in [QRL.1]. Visible credits and debits are exports (f.o.b.) and imports (f.o.b.). Invisibles are separated into services, which separately cover general government, private and public corporations. Interest, profit and dividends are similarly bifurcated. Transfers are entered by general government or private sector modes. Total invisibles are shown together with total credits and debits. The balances shown are visible balance, invisible balances in the categories above and the total invisible balance. The current balance is then given.

1.4.1 *Visible Trade: Exports and Imports*

UK Balance of Payments [QRL.61] gives trade on a balance of payments basis annually. Exports from *Overseas Trade Statistics* [QRL.48] (f.o.b.) basis are adjusted as are imports (c.i.f.). A geographical analysis is given, together with a commodity analysis.

The *Press Notice* [QRL.25] and *British Business* [QRL.7] give visible trade exports, imports, visible balance (oil, non-oil) on a balance of payments basis seasonally adjusted. The same two sources plus *Economic Trends* [QRL.30] give export and import unit values and volume index numbers (1980 = 100) as well as terms of trade (export unit value index as a percentage of import unit value index). Unit value indices are not seasonally adjusted, but volume indices are adjusted. The press release [QRL.25] gives a commodity analysis of visible trade on a balance of payments basis using basic divisions for exports, imports and visible balance: food, beverages and tobacco, basic materials, fuel, semi-manufactures, finished manufactures and total manufactures. Trade in goods other than oil is also treated in [QRL.25] and [QRL.7] on a monthly basis. Exports, imports, unit value indices are given, excluding erratics (ships, North Sea installations, aircraft, precious stones and silver) and on a seasonally adjusted basis (except unit value indices).

Visible trade in the balance of payments is given on a quarterly basis seasonally adjusted in [QRL.30]. Value of exports, imports, visible balance, oil and non-oil subdivisions are given. An annual picture of visible trade of the UK is given in [QRL.1] on a balance of payments basis: value of exports, imports, visible balance are given together with unit value index number: exports, imports, terms of trade and volume index number for exports and imports. The *Economic Trends, Annual Supplement* [QRL.30] shows the value of exports and imports with the visible balance, annually for over 25 years. It also gives annual volume and unit value indices for visible trade. A useful reference here to methodology is [B.12].

1.4.2 *The Current Balance*

The primary source for the balance of payments is the 'Pink Book' [QRL.61] also the current balance, visibles and invisibles is given in the *Press Notice* [QRL.25] and [QRL.7] on a monthly basis. The current balance, visible trade: exports, imports, visible balance (oil and non-oil) is given together with the (projected) invisible balance. Invisibles are projected and are subject to revision as more information becomes available for the most recent months. The monthly *invisibles* estimate is usually one third of the quarterly estimate (see 1.3.3 above). *Economic Trends* [QRL.30] gives visible trade: exports, imports, balance and invisibles, credits, debits, balances — total, services, IPD, transfers and the current balance on a seasonally adjusted basis. Invisibles, however, are only given quarterly except for the total balance which is given quarterly and monthly. The *Monthly Digest of Statistics* [QRL.42] gives visible trade as above, invisibles: credits, debits, balance, current balance, all seasonally adjusted, net seasonal influences and the unadjusted current balance are only given quarterly.

UK foreign trade on a balance of payments basis is given quarterly in [QRL.45]: exports, imports including and excluding erratic items (ships, North Sea oil production installations, aircraft and precious stones) volume and unit value indices for export and import (1980 = 100) and the terms of trade for all goods. Value and volume and price indices for goods and services is given on an annual basis in [QRL.61]. The value indices cover exports or imports of goods, oil, other goods, services, total goods and services. Volume indices are given for exports or imports of goods (all, non-oil), services (general government, sea transport, civil aviation, travel, financial and other, all services), total goods and services. The unit value indices are for exports, imports and terms of trade: all goods, non-oil goods. Implied price indices are provided for exports and imports of all goods, non-oil goods, services, goods and services together.

A quarterly summary of the balance of payments is given in [QRL.42]. The coverage of: visible balance, invisible balances (services, IPD, transfers, total), current balance, allocation of Special Drawing Rights, UK external assets and liabilities (transactions in liabilities, net transactions) balancing item. Quarterly summaries are also included in [QRL.30] but official financing is not broken down and the allocation of SDR's column does not include gold subscription to IMF. There are no quarterly entries in the SDR column nor in net transactions with overseas monetary authorities. Preliminary estimates for the latest quarter are included.

Quarterly summaries of the balance of payments are given in the CSO Press Notice *UK Balance of Payments* [QRL.62]. Current account items (visible trade balance, invisibile balances for services, IPD, transfers and total) and current balance are seasonally adjusted. Current balance is also given unadjusted. Investment and other capital transactions and the balancing item are given. Figures are preliminary for the latest quarter. [QRL.7] gives the same information and covers the current account details in [QRL.25]. Similar details are given in [QRL.33].

Trade in oil is given on a monthly basis in [QRL.25]. The balance of trade in oil is given as are exports and imports of oil in total and for crude oil and for the rest of SITC division 33. These figures are on a balance of payments basis and are seasonally adjusted.

Occasional articles in *British Business* [QRL.7] illuminate particular aspects of the balance of payments including invisibles e.g. earnings from Japan [B.29].

1.5 The Capital Account

The capital account is the statistical record of transactions in financial assets and liabilities between residents and non-residents. This is a measure of the flow of such transactions over a given time period. Restrictions are often drawn according to the maturity of assets. The IMF's *Balance of Payments Manual* [B.26] defines long-term capital as transactions in securities with an original term maturity of over one year or no stated maturity. Original maturity often has no bearing on the length of time for which an investment will be held so this distinction is relatively unimportant in the UK accounts (see [B.19]).

The capital account of the balance of payments covers capital transfers together with investment and other capital transactions. Prominent amongst these are: inward and outward investment (both direct and portfolio) official long-term capital, import credit, export credit, foreign currency, borrowing or lending abroad by UK banks, exchange in sterling, other external banking and money market liabilities in sterling, external lending by UK banks, other external borrowing or lending, other transactions, the balancing item and official financing liabilities and official reserves.

An overview of the balance of payments investment and other capital transactions is given in [QRL.42]. The headings used are identical to those in [QRL.33] and are derived from [QRL.61] — the definitive annual source. Overseas investment is given for UK public sector and private sector. For UK private investment overseas the following categories are listed: official long-term capital, trade credit, foreign currency borrowing or lending abroad by UK banks, exchange reserves in sterling, British government stocks, banking and money market liabilities etc., other external banking and money market liabilities in sterling, external sterling lending by UK banks, other external borrowing or lending and other transactions are listed.

The 'Pink Book' [QRL.61] gives an annual analysis of transactions in UK external assets and liabilites. Several categories of transaction are covered. Firstly, investment overseas is examined in the UK, by UK and overseas residents in both direct and portfolio forms and in total. Secondly, foreign currency lending and borrowing abroad is given, net and by UK banks. Thirdly, sterling lending is given by borrowing and deposit liabilities. Fourthly, deposits with, and lending to, banks abroad by UK non-bank private sector is given. Fifthly, borrowing from banks abroad is shown by UK non-bank private sector, by public corporations, and general government (additions to (–), drawings on (+). Sixthly, other external assets and liabilities of UK non-bank private sector and public corporations, general government and net transactions in external assets and liabilities are shown.

The 'Pink Book' [QRL.61] also gives a summary table of the levels of UK external assets and liabilities by sector at the year end, annually. Assets are subdivided by general government, public corporations, private sector and total. Liabilities are listed in the same way and net figures are similarly subdivided.

1.5.1 *Official Financing*

The concept 'official financing' used up to the 1985 'Pink Book' [QRL.61] was abandoned in 1986. Its constituent elements have been relocated elsewhere in the accounts. The official financing category brought together under one heading all transactions undertaken by or on behalf of the UK monetary authorities (HM Treasury and the Bank of England) for the purpose of reserve management and drawing on, or accruals to, the reserves themselves. The monetary authorities' use of the reserves for balance of payments purposes has declined and this category was felt to have lost its significance. Transactions previously included under this heading have been relocated elsewhere but generally remain identifiable in the accounts. The main exception is foreign currency borrowing from UK banks which is now indistinguishable from changes in banks' foreign currency liabilities. A revised presentation for *investment and other capital transactions* has consequently been adopted. Transactions from the 1986 'Pink Book' are grouped in two categories: UK external assets or UK external liabilities. The presentation of earnings (IPD) was also revised to assist association with transactions and levels.

1.6 UK Transactions within the European Community

A summary account of UK transactions with countries and institutions of the European Community is given in the 'Pink Book' [QRL.61] annually. All the major subsections of the balance of payments accounts are given. A separate table of general government transactions with the institutions of the European Community is also provided.

1.7 Balance of Payments in National Accounts Form

National income can be measured as: (1) the sum of incomes derived from economic activity by UK residents (factor incomes), (2) as the sum of expenditures or (3) as the sum of value added.

Domestic income of the United Kingdom is the income of UK residents arising from economic activity in the UK which includes the net export of goods and services. *National income* includes, in addition, income arising from investments in overseas assets including interest, profits and dividends earned abroad, but less such income earned in the UK by non-residents. This is described in the national accounts 'Blue Book' [QRL.64] as 'net property income from abroad' and is equal to the net earnings from *external assets and liabilities* in the balance of payments accounts [QRL.61]. In the international transactions table in the national accounts [QRL.64], property income paid abroad is shown both before and after deduction of UK taxes. Property income paid abroad, net of UK taxes, corresponds to earnings on UK liabilities as recorded in the 'Pink Book' [QRL.61].

Where national income is measured by totalling expenditure, final expenditure on the goods and services becoming available in the UK must include expenditure by other countries (UK exports). Many of the available goods contain imported materials and some of the available services are performed by foreign residents; expenditure on such goods and services therefore generates income and product not only in the UK economy but also in foreign ones. The gross domestic product measured as the sum of expenditure is therefore equal to total final expenditure in the UK including *exports*, on all goods and services less *imports* of goods and services. As with the income approach, an adjustment for net property income from abroad is required to convert gross domestic product (GDP) to gross national product (GNP).

The remaining items in the current account of the balance of payments, i.e. *government and private transfer payments*, do not appear directly in the composition of the national income but they are shown in the sector accounts. Government transfer payments appear in the current account of the central government. Private transfer payments appear in the analysis of personal income and expenditure.

The overall *current balance* plus *capital transfers* is described in the national accounts as 'net investment abroad'. It appears as part of the total capital expenditure in the combined capital account and since all financial transactions between different sectors of the domestic economy offset each other in aggregate, it is equal to the net acquisition of financial assets of the overseas sector in the financial accounts but is presented there with an opposite sign.

The capital transactions in the balance of payments — *transactions in UK external assets and liabilites, allocation of Special Drawing Rights* (SDRs) and *gold subscription to IMF* correspond to the financial transactions of the overseas sector in the financial accounts tables, where the analysis is broadly in terms of financial instruments [QRL.64].

The treatment of the balance of payments in the national accounts is discussed in [B.43]. A reconciliation between the capital transactions in the balance of payments accounts and the financial transactions of the overseas sector as shown in the financial accounts is included in *Financial Statistics* [QRL.33] usually in the October edition.

It is worth noting that, for the UK, the overseas sector data in the national accounts are the same as the comparable items in the 'Pink Book' [QRL.61]. The international standards: the UN system of *National Accounts* [B.45] and the *IMF Balance of Payments Manual* [B.26] would allow for differences in data published in the two contexts, if followed literally, but most countries, including the UK, put more emphasis on internal national consistency.

1.8 CSO Databank Dataset

The data in the 'Pink Book' [QRL.61] are available in computer readable form as a CSO Databank Dataset and the appropriate identifiers have been added to the 'Pink Book' tables from 1987 onwards. Details are available from the Databank Manager, CSO Branch 9, Room 52/4 Government Buildings, Great George Street, London SW1P 3AQ. The CSO does not offer direct on-line access for data, but a list of host bureaux offering such a facility is available from the CSO.

CHAPTER 2

UNITED KINGDOM VISIBLE TRADE

2.1 Introduction

Visible trade covers imports and exports of goods or merchandise i.e. in principle their ownership changes in the course of the transaction between a United Kingdom resident and a foreign resident. Visible trade in goods is contrasted with trade in services 'invisibles' which are covered in the following chapter. In principle, a change of ownership occurs even when the goods are transferred between units which a firm owns abroad.

In examining trade statistics it is important to distinguish between the raw data and the information as presented in official statistics. The raw data are collected by HM Customs and Excise in Southend-on-Sea from documents completed by importers and exporters. An unofficial but authoritative source estimates that 40 per cent of documents going to Customs and Excise at Southend-on-Sea contain errors. These raw data are examined in sections 2.2, 2.3 and 2.7 below. This information is then compiled from Custom and Excise records into *Overseas Trade Statistics* [QRL.48] monthly and [QRL.49] annually, (sections 2.4, 2.5, 2.7, 2.8). The figures from *Overseas Trade Statistics* are amended so as to conform with the balance of payments basis as described in 1.2 above, and visible trade figures on a *balance of payments basis* appear in the 'Pink Book' [QRL.61].

The subgroupings of international trade statistics have been agreed internationally in the UN code known as the Standard International Trade Classification (SITC). This code is a major help in making the statistics of individual nations comparable.

The composition and direction of trade are important both for government policy and for individual firms, in particular in identifying their 'key markets' and trends in those markets.

It is particularly important to be able to identify the areas of the economy which are most able to withstand foreign competition both in the home market and in foreign markets. A great deal of recent attention has been paid to the international competitiveness of British industry and the construction of measures of competitiveness at sector, industry and firm level is of great importance. In the context of debates on de-industrialisation and the increased 'globalisation' of competition, the competitiveness debate has centered on manufacturing industry and it is here that the search for comparators, for competitiveness is a relative concept, is most crucial (see 2.12 below).

2.2 Definition

Figures on visible trade cover exports from and imports into the UK of goods, whether or not intended for use in the UK or for re-export. Goods are recorded as they cross the national (geographical) frontier of the UK rather than the Customs boundary; this system is known as the 'general' system of recording trade. In principle only those transactions involving a genuine exchange of resources are counted; a transfer of ownership between UK and foreign residents being a necessary condition.

For EC needs only, certain figures are prepared on a 'special' trade basis. Under this system of recording, the statistics of imports and exports are restricted to those goods actually passing through the Customs boundary.

2.3 Classification

2.3.1 *Classification pre-1988*
Statistics of the UK's overseas trade are collected by HM Customs and Excise within the framework of the CCCN (Customs Cooperation Council Nomenclature), an internationally agreed classification for customs purposes, previously called the BTN (Brussels Tariff Nomenclature). The Department of Trade and Industry publishes the statistics of overseas trade according to the headings of the UN Standard International Trade Classification (SITC). From January 1978 the headings used are those of the enlarged and revised version (SITC (Rev2)) published by the UN in 1975. The headings of the SITC (Rev2) and the SITC (R) which was used from 1960-77 correlate with those of the CCN but are in a different order. The headings against which trade statistics are collected are listed in the Annual *HM Customs and Excise Tariff and Overseas Trade Classification* [B.39] which is used for both duty purposes and trade statistics. Since 1st January 1978 a single integrated classification (HM Customs and Excise Tariff and Overseas Trade Classification) has been used. *The Guide to Classification for Overseas Trade Statistics* [B.38], published annually, lists the statistical headings that appear in the tariff together with the eight-digit code numbers under the appropriate headings of the SITC (Rev2).

The Tariff is divided into the 21 Sections and 99 Chapters of the CCCN which is in turn split into 1095 headings identified by four digit code numbers. Following membership of the EC, the UK Tariff also incorporates the duty headings of the EC's Common Customs Tariff (CCT) which comprises further subdivisions of the CCCN. Also, since 1978, the UK has incorporated into its trade classification the statistical headings of the EC classification NIMEXE (Nomenclature of Goods for External Trade Statistics of the EC and Statistics of Trade between Member States). Some of these headings are subdivided further for national statistical purposes and these sub-divisions are indicated by positive numbers as the seventh and eighth digits. Each statistical heading in the Tariff is identified by a unique eight digit code number of which the first four digits indicate the CCCN heading under which it falls and the first six digits identify the NIMEXE heading.

2.3.2 *Classification post-1988*

From 1st January 1988 the UK introduced a new system of goods classification based on the worldwide Harmonised System. In the early 1970s the need to develop a single goods description system to satisfy all the requirements of international trade was recognised. The increased use of computers meant that a common coding system was desirable. As a result Customs and Statistical Administrations co-operated with international trade and transport bodies to develop a Harmonised Commodity Description and Coding System (HS) which was finally adopted on 14 June 1983 (the EC signed the Harmonised System Convention in June 1985, ratification by sufficient countries needed to bring the Convention into force on 1 January 1988 occurred in September 1987). It is expected that the majority of world trade will be reported under the Harmonised System thus facilitating international comparisons of statistics.

The new HS is broadly comparable with the CCCN but is more detailed and up-to-date in both structure and content. It meets the objectives of an international goods coding system not only for Customs tariff (duty etc.) purposes but also for freight and commercial uses. The new headings are correlated to SITC Revision 3, which has been constructed by the UN Statistical Office and will be introduced by the UK at the same time as the HS.

The new UK Tariff combines the EC's HS-based duty and statistical nomenclatures, within which further sub-divisions for national statistical purposes, have also been incorporated. The most obvious effect of this change has been to replace the previous eight digit Tariff Trade Code Number with a nine digit Commodity Code.

2.3.3 *Erratics*

In the context of short-term statistics certain terms can be considered erratic. These items are trade in ships, North Sea production installations, aircraft, precious stones and silver bullion (SNAPS). Precious stones are classified to 'non metallic mineral manufactures' (Division 66), silver bullion to 'non ferrous metals' (Division 68) and other erratic terms to 'other transport equipment' (Division 79).

2.3.4 *Transit Trade*

Goods in transit through the United Kingdom are not included in the overseas trade statistics.

2.3.5 *Parcel Post*

Imports and exports by parcel post are, in general included in SITC Rev 2 sub group 911.0 (Postal packages not classified according to kind). Figures given are therefore deficient to the extent of the volume of parcel-post trade. This does not apply to precious and semi-precious stones exported or imported by post which are included in the figures for the relevant SITC. As a general rule, packets sent by 'Letter Post' or 'Printed Paper Post' are not included in the statistics.

The value of parcel post trade is estimated by applying average value of parcels to the number of parcels. These average values are derived from occasional sample surveys extrapolated by appropriate unit value indices.

2.3.6 *Low Value Consignments*
Imports and exports (other than in 2.3.5 above), other than imports subject to excise duty of an individual value less than a given amount (£200 for 1987) are aggregated under SITC (Rev 2) sub group 931.0, 'special transactions and commodities not according to kind' as 'Transactions of low value' against 8 digit code 9999 0299. This trade is not analysed either by commodity or country. Figures for individual country and commodity headings are therefore deficient to the extent of these consignments, which amount to a very small proportion of total trade.

2.3.7 *Exclusions*
Certain classes of goods are excluded from the figures in [QRL.49]. These include temporary importations or exportation, personal and household effects and legacies, sea-going vessels sold abroad, stores and fuel etc. Where individual categories of commodity are being examined, care should be taken to establish that the exclusions listed in [QRL.49] are unlikely to be significant.

2.3.8 *Additional Information*
As mentioned above, the original records, from which overseas trade statistics are compiled, comprise information on each of the commodity headings, numbering about 9000, which make up the Tariff. Details of the trade descriptions are published in the *Guide to the Classification for Overseas Trade Statistics* [B.38] arranged in the order of the SITC (Rev 2).

Information for any of the commodity headings and for the 215 areas and countries distinguished and also for trade through United Kingdom ports may normally be obtained (if commercial privacy does not prevent disclosure), either singly or on a regular monthly or cumulative basis on payment of a fee [B.28]. These analyses are now available from Marketing Agents whose names and addresses may be obtained from the Statistical Office, HM Customs and Excise at Southend-on-Sea, Essex.

2.3.9 *Terminology*
The International Chamber of Commerce produces *Incoterms* (International Rules for the Interpretation of Trade Terms) on an irregular basis, the latest being *Incoterms 1980* (No. 350) [QRL.35]. *Incoterms* gives codifications of terms already accepted by long usage and represents a codification of current practices awaiting the same degree of acceptance. The terms have no formal legal status, although they are widely accepted in legal and commercial circles. *Incoterms* is currently undergoing revision. The main use of *Incoterms* is for businesses to use the most appropriate terms of delivery.

2.4 Direction of Trade

The regional breakdown of UK trade with the rest of the world is given for individual countries and for areas of the world classified as: the European Community, Rest of Western Europe, North America, Other Developed Countries, Latin America, Middle East and North Africa, Other Developing Countries, Centrally Planned Economies, EFTA, OECD, Oil Exporters and the Commonwealth. They are given annually in *Overseas Trade Statistics of the United Kingdom* [QRL.49]. These figures are also given for the latest month and for the total of months in the year ending at the latest month, [QRL.48] (see also [QRL.43]). The value of exports and imports by areas of the world is also reported in [QRL.7], [QRL.42] and [QRL.1].

The *Monthly Review of External Trade Statistics* [QRL.43] contains information on UK visible trade as well as some international and miscellaneous data. It is prepared as an internal working document but is published by the Department of Trade and Industry on a regular basis. It contains three main sections: a commentary section on the latest monthly trade statistics, a section containing graphs of key statistical series and a main section tables. The tables cover (a) UK trade on a balance-of-payments basis (b) UK exports on Overseas-Trade-Statistics basis (c) UK imports on Overseas-Trade-Statistics basis (d) derived statistics of UK trade (e) international statistics and (f) miscellaneous statistics. In addition the annual supplement [QRL.44] gives longer historic runs of data, if available, for all series in the monthly edition, together with certain other series which cannot be updated regularly. A mid-year supplement is also published.

The derived statistics of UK trade cover (1) crude balances and relative values by major commodity group (2) crude balances and relative values by major area group and (3) relative volume and terms of trade by major commodity group.

International comparisons compare (1) the UK and the main manufacturing countries' (MMCs') exports of manufactured goods which gives the UK share of the total together with a volume index for MMCs' and UK and (2) the individual MMCs' exports of manufactured goods in percentage share terms. These figures are given annually for the past eight full years and quarterly for the past six full years plus the current year. Main manufacturing countries are defined as: UK, Federal Republic of Germany, Belgium/Luxembourg, France, Italy, Netherlands, Sweden, Switzerland, USA, Canada and Japan.

Miscellaneous statistics cover: sterling exchange rates, MMCs' exports of manufactures-unit value indices, UK competitiveness in manufactured goods, unit labour costs, normalised unit labour costs, relative unit labour costs, relative normalised unit labour costs, relative export unit value indices, export orders by industry, summary ECGD statistics and invisible exports and third country trade insured by ECGD [QRL.43].

The *Annual Supplement* [QRL.44] also covers the share of exports by major commodity group, the share of imports by major commodity group, the share of exports by major area group and the share of imports by major area group, all on an overseas trade statistics basis. In international comparisons, the *Annual Supplement* also covers the UK shares of imports of manufactures to the major markets and comparisons of the UK import shares with its major competitors. The

UK share of OECD exports to major markets are also given. The supplement also shows the UK's top 50 export markets for the most recent year and 10 years previously and the top 50 UK import sources for the same years.

2.5 Commodity Composition of Trade

The breakdown of the value of imports and exports by commodity division and by area of the world and countries is also available on a monthly basis [QRL.48]. Areas are as above and the Commodities are divided by SITC (Rev 2) divisions and sections (i.e. 1 and 2 digit code numbers). A more dissaggregated breakdown occurs where commodities are divided by SITC groups, subgroups and items (i.e. 3, 4 and 5 digit code numbers) in total and for European Community and any countries accounting for significant amounts. Monthly figures are given as in section 1.2. The finest breakdown by commodity category of UK trade is given where the breakdown is by SITC and 8 digit code numbers for every trade description in the tariff and overseas trade clasification. The Commodity Headings are in SITC Subgroup or Item Order. Headings are regularly changed and it is often necessary to introduce new trade descriptions in order to take account of changes in the basic tariff structure. Both the original and replacement 8 digit code numbers are included. They are identified by a letter symbol immediately following the value figure. These data are given for the latest month and the period ending in that month, [QRL.43] (see also [QRL.7] [QRL.42] [QRL.1]).

2.5.1 *Area x Commodity Analysis*

An annual analysis of UK external trade by area and commodity is provided in [QRL.63] which appears quarterly. Country x commodity entries are given for (1) total exports (f.o.b.), (2) total imports (c.i.f.), (3) exports less imports and (4) exports over imports percentage of commodities by SITC section and Division Headings (1 or 2 digit code numbers). All European countries are listed separately as are Japan and the USA. The areas covered are EC, rest of Western Europe, EFTA, total Western Europe, OECD, oil exporters, Latin America, Centrally Planned Economies and world total.

The export:import ratios (percentages) are included because crude balances can give a misleading indication of relative trading performance over time because of the impact of inflation on the levels of trade. However, it is not clear how expression as a percentage corrects for price changes over time, (see also [QRL.48]).

An analysis of 'UK Visible Trade in the Post-War Years' is available in [B.14].

2.5.2 *Commodity Flow Accounts*

Commodity Flow Accounts, compiled by the CSO [QRL.21] provide for 43 different commodity groups, the proportion of total output going to exports and coming from imports. This is the only information available on an input-output basis. The figures are less detailed than the UK input-output tables, but much more up-to-date. The Commodity Flow Accounts are compiled by the CSO but are available only where payment is made. For full details see Astin and Brady [B.3].

2.5.3 *Special Chemical Statistics*

Special Chemical Statistics Ltd provide details on a monthly basis of the quantity, value and country of consignment of imports to the UK of the majority of single organic chemicals [QRL.57]. The fully computerised return provides a Special Chemical Return in order of Tariff Trade Code Number (TTCN) with the products grouped under each heading. Subscribers are provided with an alphabetical list of precise chemical names of products giving the appropriate TTCN numbers as a key. An annual fee is payable to the private company which markets the return.

2.5.4 *Correspondence Between SITC (R) and SIC*

Although exports and imports are available by industry (i.e. SIC classification) in *Business Monitor* MQ10 [QRL.15] no published correspondence between SITC (R) and SIC currently exists. An unofficial 'assumed correspondence' is provided by Kirsty Hughes [B.11] p. 178-9 based on an earlier attempt by Armstrong [B.1]. This aligns Minimum List Headings with three-digit SITC(R) codes.

A computer-based formula transforms customs data from SITC to SIC for the purpose of constructing MQ10 [QRL.15]. It would be a major boon if future classifications were made crystal clear by the publication and updating of a correspondence between the most important trade classification and the most important domestic industrial classification.

2.6 Gold and Monetary Items

Gold and monetary items have been subject to changes in treatment over the past years. Since January 1981, all trade in gold, other than that in unwrought refined bullion is included in merchandise trade. Imports and exports of monetary gold are shown separately in the 'Pink Book' [QRL.61].

2.7 Timing and Correction

In general the figures show the trade as declared by importers and exporters on the completed documents received in HM Customs and Excise Statistical Office at Southend-on-Sea during a month.

Under the procedures for the control of exports, goods may be cleared for export using a simplified entry and traders are then allowed fourteen days from the date of shipment to provide a complete export document. Moreover, the processing of export documents begins shortly before the end of the calendar month. Thus the export statistics compiled for a month do not correspond with goods shipped in that calendar month.

Importers must usually present complete documentation to obtain Customs clearance and remove their goods. Further, the import figures include documents reaching Southend-on-Sea up to the third working day after the end of the month. As the majority of import documents are in fact transferred in the form of magnetic tapes overnight after being cleared, delays are negligible and import figures correspond fairly closely to goods actually imported during the calendar month.

Figures are subject to correction. For example, goods such as fruit and vegetables imported on consignment may be declared on a provisional value which is later amended. Errors may also arise from mis-classification by commodity in the original document, or in respect of valuation; and they may arise from the non-receipt of a statistical document, or the submission of more than one document for the same consignment.

Corrections relating to earlier months, if they are significant, are incorporated in the cumulative totals only. Consequently, cumulative totals may differ from the sum of monthly figures. The annual edition of *Overseas Trade Statistics of the UK* [QRL.49] contains updated information taking into account corrections notified to HM Customs and Excise after the monthly figures were compiled. The basis for seasonal adjustment to the overseas trade statistics is explained by Elis-Williams [B.8].

2.7.1 *Reliability of the Overseas Trade Statistics*

The Customs and Excise Statistical Office devotes a large part of its resources to ensuring that the Overseas Trade Statistics are of sufficient accuracy to satisfy users. During routine monthly processing, documents are subjected to editing: those failing checks are presented for investigation and possible correction. In addition, a number of special exercises are conducted to monitor the accuracy of the statistics. In the case of exports there has been a publicised problem of under-recording, caused by exporters failing to submit complete documentation after using the simplified procedure, and of overvaluation. Regular estimates of these two offsetting effects are obtained and adjustments made to balance of payments aggregates as necessary. Nevertheless both total exports and total imports on an OTS basis are considered to be of good reliability in recent years; broad commodity and country breakdowns also justify this rating.

The quality of the detailed Overseas Trade Statistics is harder to assess. Regular sample analyses, which compare the information used in the compilation of the statistics with the 'true' position obtained by examining commercial documents, show that some degree of misclassification between headings occurs at the 5th digit (Tariff Trade Code Numbers) and below. However the position is not uniform as data editing is concentrated on those headings which users have indicated as important. Where accuracy at the Tariff Trade Code Number or Commodity Code level is crucial advice should be sought from the Customs and Excise Statistical Office.

2.8 Valuation of Trade

The basis of valuation in the *Overseas Trade Statistics of the UK* [QRL.48], [QRL.49] is information given for imports under paragraph 17 of Part 3A of Volume II of the Tariff and for exports under paragraph 16 (Box 27) of Part 3B. The present valuation code came into force on 1st July 1980.

Imports. For goods subject to duty *ad valorem* the statistical value is the same as the value for Customs purposes. The value is arrived at by specified valuations which have to be applied in strict hierarchical order. These are, in order, the *transaction value* of the imported goods (broadly the price paid or payable for those goods), the transaction value of identical goods, the transaction value of similar goods, the deductive method (value derived from the selling price of the goods in the country of importation) and computed value based on the built-up cost of the imported goods. (The order of the last two can be reversed at the importer's request). Finally a fall back provision for exceptional cases is applied.

The 'place' for valuation purposes is the point where the imported goods are introduced into the customs territory of the European Community. This means that the costs for delivery of the imported goods to that place have to be included. From 1st February 1982, for conversion to sterling of an amount expressed in foreign currency, a system of 'period rates' of exchange is applied. This (from October 1985) is a weekly rate, commencing each Wednesday although there is provision for the rate to be changed in exceptional circumstances.

For all other goods, i.e. goods free or exempt from duty and goods subject to a specific duty, the statistical value is determined in relation to the point at which the goods enter the UK. This means that the value includes freight, insurance and all other costs, charges and expenses incidental to the sale and delivery of the goods to the port or place of importation into the UK with the exception of any duty or tax chargeable in the UK. When the goods are re-imported after process or repair abroad the value includes the cost of the process or repair and the value of the goods when exported.

Exports. Export values are on an f.o.b. (free on board) basis i.e. the cost of the goods to the purchaser abroad, including packaging, inland and costal transport in the UK, dock dues, loading charges, and all other costs, profits, charges and expenses (e.g. insurance) accruing up to the point where the goods are deposited on board the exporting vessel or aircraft or at the land boundary of Northern Ireland.

In general, import valuations are of better quality than export valuations. Import valuations from outside the EC are of better quality than from inside. When a duty is involved then a better valuation is given.

2.9 Cross-Classification: Invoicing Currency

British Business [QRL.7] gives the invoicing currency of exports and imports by area, commodity and size of transaction on a biannual basis.

2.10 Index Numbers: Unit Values and Volumes

Commodity Export Volume Indices (1980 = 100) by major commodity group (and excluding erratics), non-manufactured goods, semi-manufactured goods, finished manufactured goods (and excluding erratics) based on broad economic categories and/or SITC are given, together with unit value indices in *Monthly Review of External Trade Statistics* [QRL.43]. This is the primary source. Export and Import Unit Value Index Numbers are given by commodity classes based on SITC section or division and UN classification by broad economic categories (BEC) on a monthly basis (1980 = 100) in *Monthly Digest of Statistics* [QRL.42].

The value of commodity imports, commodity imports volume and unit value indices (1980 = 100) are broken down by major commodity groups based on SITC and excluding erratics on a monthly basis in *Monthly Review of External Trade Statistics* [QRL.43] seasonally adjusted (except for unit value indices). This also gives a breakdown of the value of imports and imports volume and unit value indices for food, beverages, tobacco and basic materials (1980 = 100) on a monthly, similarly seasonally adjusted basis. Also analysed are semi-manufactured and finished manufactured (also excluding erratics) goods.

An *Annual Supplement* [QRL.44] gives longer historic runs of data, as available for all series covered in the *Monthly Review of External Trade Statistics* [QRL.43]. The 'Pink Book' [QRL.61] provides value and volume and price indices for goods and services annually. For each year, quantities of trade in individual goods and services are revalued using the prices ruling in a given 'base' year. For a single good, the choice of base year is immaterial and the series will reflect accurately volume changes.

However, when two or more revalued series are added together, the weights given to the volume changes of the various components will be determined by their prices in the base year and changes in relative prices lead to these weights becoming out-of-date. Switches to base years are therefore made and aggregate revalued series are rescaled onto later base year prices using selected 'link' years. For 70 per cent of trade in goods the current value of trade is divided by the ratio of its current price to its price in the base year. (Primary data are from *Overseas Trade Statistics* [QRL.49]). Where this cannot be done the price is 'assumed to have moved in line with a similar commodity, or group of commodities, for which satisfactory estimates can be made' (Pink Book). This estimation procedure is used for a further 20 per cent of trade, for the remaining 10 per cent price movements are 'guesstimated' from other sources such as producer price increases. For trade in services, it is usual that only values, not volume is known. For imports an index is derived from price indices of the exporting countries weighted together and converted to sterling. For many exports of services a UK price index is issued. These procedures give rise to severe doubts on the possibility of accuracy of volume indices for services. This difficulty results in regular revaluations of service trade and is very disconcerting for balance of payments watchers and forecasters.

2.10.1 *The Terms of Trade*

The terms of trade of a country are the price of its exports relative to the price of its imports. Developments in the terms of trade are analysed in [B.4] although this article is dated September 1980.

2.11 Import Penetration and Export Sales Ratios

Business Monitor MQ12 [QRL.16] gives import penetration and export sales ratios for manufacturing industry on a quarterly basis. This information covers the 12 months ending in the relevant quarter and covers the past 2 years with an extra entry for the provisional estimates for the latest quarter.

[QRL.16] presents four ratios relating overseas trade in the goods produced by virtually every sector of manufacturing to domestic production and apparent consumption of those goods. The ratios were first introduced in the August 1977 edition of *Economic Trends* in an article 'The Home and Export Performance of United Kingdom Industries' which described the conceptual and methodological problems involved in measuring such variables as 'import penetration' [B.25]. Annual ratios at MLH level for the years 1970 to 1979 were presented in an article 'Manufacturing Industry in the Seventies: an Assessment of Import Penetration and Export Performance' in the June 1980 edition of *Economic Trends* [B.10].

In the *Monitor* [QRL.16] the latest ratios are supplemented with provisional estimates of the ratios at class level and above for the next period.

The ratios relate to twelve month periods at quarterly intervals. They are expressed as percentages and defined as follows:

1. Imports: home demand (where home demand = manufacturers' sales + imports – exports)
2. Imports: home demand + exports
3. Exports: manufacturers' sales
4. Exports: manufactuers' sales + imports

The sales figures used in the preparation of these ratios relate to UK manufacturers' sales of the principal products of the industry concerned excluding, where possible, waste products and work done and including estimates for the sales of firms too small to be included in the enquiries. Generally, these are derived from the appropriate *Business Monitors* whilst the overseas trade series used are based on the figures compiled for *Business Monitor* MQ10 'Overseas Trade Analysed in Terms of Industries', [QRL.15]. Imports are valued c.i.f. and exports f.o.b. The provisional estimates are calculated in the usual way but use early estimates of the sales figures. All data are in current prices.

These figures are repeated in [QRL.7], [QRL.42] and annually in [QRL.1].

2.12 Competitiveness

Measures of UK competitiveness in trade in manufactures are given in [QRL.30]. Summary measures are: (1) relative export prices, (2) relative producer prices (3) IMF index of relative unit labour costs (actual and normalised), (4) import price competitiveness, (5) relative profitability of exports. These are given on a quarterly basis. The export unit value index, producer price index and unit labour cost index are given for the US, UK, Japan, France and the Federal Republic of Germany, (see also [QRL.42]).

Summary measures of competitiveness are described in *Economic Trends* No 304, February 1979 [B.7] and revised measures of import price competitiveness and relative profitability of exports are described in [B.36] *Economic Trends* No 319, May 1980, 'Revised Measures of UK Trade Competitiveness in Manufactures: Technical Note'.

2.13 Scotland

The value of imports and exports through Scottish ports by commodity classes based on SITC Division and Section titles is given on an annual basis in the normally biannual *Scottish Economic Bulletin* [QRL.56]. The Scottish Council for Development and Industry has produced estimates of Scotland's manufactured exports by SIC coding and by foreign market, annually since 1984 (triennially before this date) [QRL.55].

2.14 Northern Ireland

Import and export figures for Northern Ireland are available up to the end of 1973 by value and commodity. A series on tonnage of goods imported and exported through principal ports ends in 1980. The source of these two defunct series is *Northern Ireland Annual Abstract of Statistics* [QRL.46]. Additional data were given in *Statistics of Trade through UK Ports* which gave trade analysed by commodity and port of importation or exportation but this published data ceased in 1980. From 1980 onwards there are no published data on imports and exports for Northern Ireland separately.

Data on travel and tourism are however available [QRL.46]. Total numbers of passengers and passenger-accompanied cars leaving Northern Ireland by sea and air are given, including both visitors and Northern Ireland residents. Home holidays are given by long holiday (4 + nights) and short holidays (1-3 nights) with associated spending. Total home-holiday revenue is also given. Total tourism (trips, nights, spending) is presented by district council area for home holidays and staying visitors together. Tourism is analysed annually by staying visitors (total numbers, total revenue) by origin (from Great Britain, Irish Republic, Overseas) by purpose (visiting friends and relatives, holiday, shopping, business other). Day excursions are also given (total visitors and total revenue). The source of this data is the Northern Ireland Tourist Board.

Details of the are of inward investment to Northern Ireland (number of jobs, cost per job, source nation) are available in *Industrial Development Board Annual Reports* [QRL.36] which deal with firms with 50 or more employees and *Northern Ireland Local Enterprise Development Unit Annual Reports* [QRL.41] deal with firms of under 50 employees. Other regional data on other measures of inward investment (assets etc.) can be derived from the UK Census of Production [QRL.52] (see also section 5.6 below). The *Northern Ireland Annual Abstract of Statistics* [QRL.46] includes jobs promoted by the Industrial Development Board (and its predecessor agencies) by new companies, new companies within Northern Ireland and expanding companies by year and for the most recent two years by quarter. Jobs promoted by the Local Enterprise Development Unit are given by year and for the most recent three years by quarter (initial contract companies, further contract companies) [QRL.46].

2.15 Custom Duties

Custom duties are given in the annual *Report of the Commissioners of HM Customs and Excise* [QRL.54]. Total imports are analysed by SITC (Rev 2) Section 1 digit code numbers showing dutiable status and duty charged for imports from the whole world and from the EC.

2.16 Implications of the European Internal Market

The EC's requirements for overseas trade statistics are becoming an increasingly important factor for the Statistical Office of HM Customs and Excise [B.9]. The substantial range of statistics and surveillance information provided to the EC has already been described. However, the proposals in the EC 'White Paper'—*Completing the Internal Market* [B.33] could have a much greater impact on the compilation of trade statistics than anything that has gone before. Internal market developments and their effect on the compilation and availability of intra-community trade statistics, are likely to be major concerns in Statistical Office planning in the next five years.

CHAPTER 3

UNITED KINGDOM INVISIBLE TRADE

The *UK Balance of Payments* [QRL.61] divides UK invisible trade into four main sections: services, earnings on UK external assets and liabilities (interest, profits and dividends), overseas earnings of UK financial institutions (The City) and transfers. This chapter follows that order after a section giving an overview of UK invisible trade.

The reliability of estimates of invisible trade was discussed in detail in section 1.3.3.

3.1 Overview

In the summary balance of payments, invisibles are given as service balance; interest, profits and dividend balance; transfers balance; and invisibles balance being a summary of the three sub-balances [QRL.61]. The current account summary tables give both credits and debits for: services (general government, then private sector and public corporations subdivided by sea transport, civil aviation, travel, financial and other services) interest, profits and dividends (general government, public sector and private corporations), transfers (general government and private sector) and total invisibles.

An overview of world invisible trade is given by the British Invisible Exports Council in *World Invisible Trade* [QRL.67]. This is the only regular analysis of world invisible trade. The first report appeared in 1967 (1964 data) and subsequent reports have been published in 1969, 1971 and each successive year since then. The report shows total invisible receipts (excluding miscellaneous government receipts) in US $ by country for the top 35 countries (list attached as Appendix 4) and for the rest of the world. Growth rates are also shown in percentage terms. Invisible trade as a proportion of total trade is given and this proportion is shown for receipts and payments by country for the top 35 countries and for the rest of the world. Invisible trade is further shown as a proportion of GNP. The main categories of invisible trade: transport, travel, investment income, other services are shown by country (for the top 35 countries), for the rest of the world and in world total figures. National balances by main category and by country are shown for total receipts and payments and a growth index of total invisibles (1969 = 100) is compiled.

Quarterly estimates of the UK balance of payments including invisibles are published by *Press Notice* each quarter [QRL.62] and details are included with the Press Notice in March, June, September and December issues of *Economic Trends* [QRL.30], the September issue figures being consistent with those in the 'Pink Book'.

The latest quarterly estimates are also reproduced in summary form in the *Monthly Digest of Statistics* [QRL.42].

Monthly estimates of visible trade and the projected invisible balance, expressed as a monthly rate though not based on monthly data, seasonally adjusted are published monthly in [QRL.25] and subsequently in *British Business* [QRL.7]. Details of the compilation of the monthly invisibles balance is given in [B.35], see also section 1.4.

3.2 Invisibles: Services

The 'Pink Book' [QRL.61] distinguishes the following categories of services:

1 General government — all UK government transactions not included elsewhere, including those of its employees stationed abroad
2 Sea transport and civil aviation — the carriage of passengers and freight, the chartering and hiring out of ships and aircraft, port services and the provision of fuel and other goods and services associated with the operations of ships and aircraft
3 Travel — goods and services provided to UK residents while they are abroad and to foreign residents in the UK
4 Financial and other services — services provided by UK financial institutions and consultants; other services provided to and by UK companies; services paid for by royalties; overseas students' and governments' expenditure in the UK not included elsewhere and other miscellaneous services.

Categories 2 to 4 are services of private sector and public corporations.

3.2.1 *Valuation of Services and Coverage*

Most services are valued by the payment made, information being collected from either the exporter or the importer of the service (see 1.3.3).

Visible trade estimates cover imports valued f.o.b. (i.e at the exporting countries' ports) but less the value of loss and damage incurred *en route* to the UK. For UK importers who purchase goods f.o.b. they bear the cost of the value of loss and damage incurred *en route*, plus freight services outside the exporting country and insurance services (the excess of insurance premiums paid for the journeys over the claims made). These are debit entries. A similar reasoning applies to costs of freight and insurance paid by overseas importers. Such services provided by UK residents give rise to credit entries. Analagous debit and credit entries under 'land transport' occur when the foreign country operators carry the goods within the importing country.

The trade statistics include the full values of UK imports, even where the actual payment is reduced to take account of services provided by UK importing agents. The commission of these agents is therefore included as a credit entry under 'other services'. Similarly, the commission of non-resident agents dealing with UK exports are included as a debit entry.

3.2.2 *General Government*

In the balance of payments invisible account [QRL.61], general government services include:

(1) all exports and imports of services by the UK government, including its employees stationed abroad, to or from overseas residents, unless these transactions are more appropriate to other items in the account (there is thus an arbitrary element in the classification);

(2) imports of goods from non-UK sources to UK government installations abroad (goods which have passed through HM Customs and Excise are excluded as they are UK imports in the visible trade accounts). There figures are obtained from official sources including returns provided by governmental departments and are therefore considered accurate.

Credits are in respect of European Community institutions, US forces expenditures and other military receipts by UK governmental and other receipts. Debits are military expenditure, administrative, diplomatic and other expenditure.

3.2.3 *Sea Transport*

Exports of UK operators are freight services on UK exports and on cross trades (trade between countries other than the UK), the carriage of overseas passengers and the provision to them of goods and services, and the chartering of ships to non-residents. Exports also include fuel and other supplies purchased in the UK by overseas operators. Exports include fuel and other goods purchased abroad by UK operators, their chartering of ships from non-residents and the carriage by non-resident operators of UK imports and UK passengers.

UK Balance of Payments [QRL.61] gives a detailed annual breakdown of credits, debits and balances of payments for sea transport services. Credits are for ships owned or chartered-in by UK residents and disbursements in the UK relating to ships operated by overseas residents. The first category is sub-divided by operating earnings and time charter receipts for both dry cargo, including passenger revenue, and wet cargo. Debits are analagous. The quarterly *Statistical Brief of the General Council of British Shipping* [QRL.58] gives UK balance of payments figures for UK-owned ships by year of contribution. Gross earnings are broken down into gross earnings from abroad and gross import savings, the latter being freight on imports, passage money collected in the UK and time-charter receipts from non-UK shipowners. UK balance of payments contributions are also divided by ship type (liner, tramp and tanker) in total and that from cross trades. A geographical analysis of UK shipping from the DTI and General Council of British Shipping is also available [B.31]. A full review of UK statistics on sea transport is given in [B.16].

Business Monitor MA8 'Nationality of Vessels in United Kingdom Seaborne Trade' [QRL.12] covers the seaborne trade of the UK as recorded by HM Customs and Excise. From 1988 the publication will be produced by the Department of Transport. More detailed statistics can be purchased from the Department of Transport. Analyses of the earnings of the UK shipping industry from its international activities were previously published in *British Business* [QRL.7], but are now published as Department of Transport "Statistical Bulletins". The second in this series is *United Kingdom Shipping Industry: International Revenue and Expenditure*

1985 [QRL.65]. *Business Monitor MR15* 'General Trends in Shipping' [B.32] gives details on both the UK fleet and world fleets together with a commentary. As from 1987 it too will be published as a "Statistics Bulletin" by the Department of Transport.

Seaborne trade is covered in [QRL.12]. It gives eleven-year runs of information on weight of cargo carried and percentage by flag, tonne-miles of cargo carried and percentage by flag and value of cargo and percentage by flag. Three-year runs of information are given on weight of cargo imported (and, separately exported) and percentage by flag to and from principal countries and areas. For the most recent year, tonne-miles of cargo carried and percentage by flag is subdivided by short and deep-sea routes. The average length of haul by flag and cargo types is given for the most recent year. The proportion of trade carried by principal flags is shown for the past three years. A three-year analysis is given of weight of cargo imported and percentage by UK flag by major commodities and countries from which consigned. Similarly, the weight of cargo exported and percentage by UK flag is analysed by major commodities and countries to which destined.

3.2.4 *Civil Aviation*

A similar rubric applies to the import and export of civil aviation services as to sea transport. The transactions of UK airlines are derived from returns supplied by the airlines. On credits, UK airlines' passenger revenue relates to all tickets sold outside the UK and used on British aircraft together with excess baggage receipts. It is assumed that most tickets sold abroad are sold to overseas residents and that the small number actually sold overseas to UK residents is counter-balanced by sales in the UK to overseas residents (which are not measured). In so far as these assumptions do not hold, a source of error is introduced. On debits for UK airlines, these include charges and purchases abroad of goods and services. Overseas airlines' credit entries are expenditures in the UK, based on information from oil companies on supplies to overseas airlines, returns from the British Airports Authority and other airports for landing fees and other airport charges and information supplied to the Civil Aviation Authority by the majority of large overseas airlines operating in the UK on other UK expenditures. Debits for overseas airlines are derived from the International Passenger Survey (IPS) on fares. To this is added estimates of expenditure of UK residents on flights to Ireland which the IPS does not cover. Up to the 1986 'Pink Book' [QRL.61] only payment for flights immediately to and from the UK was covered. These figures have been adjusted in the 1987 'Pink Book', but not to earlier years, producing a break in the series. Estimates of overseas' airlines' freight in UK imports are derived by subtracting from the estimates of total freight on imports arriving by air, estimates of the element provided by UK airlines. Other debits are payments to overseas airlines for carrying UK airmail and this is reported by the Post Office to the Civil Aviation Authority.

The number of separate sources, estimates and strong assumptions in the data on civil aviation gives rise to concern about the accuracy of the reported figures which, therefore, should be used and interpreted with caution.

The balance of payments credits, debits and balances for civil aviation are given annually [QRL.61]. The usual quarterly and monthly balance of payments estimates are also available. Credits are sub-divided by UK airlines: overseas passenger revenue, freight on UK exports and cross-trades and other. Overseas airlines disbursements in UK make up the total. Debits are for UK airlines disbursements overseas in total and for overseas airlines broken down by UK passenger revenue, freight on UK imports and other. Balances are given for UK and overseas airlines and an overall total balance for civil aviation. An analysis of the UK air transport industry is given in [B.23].

Economic Trends [QRL.30] gives information on civil aviation, 'British Airports and Traffic Statistics'. Information on the *Air Transport Industry* is analysed in [B.6] and a full review of UK statistical sources on civil aviation is given by Celia Phillips [B.20]. The Civil Aviation Authority publishes information in *CAA Monthly Statistics* and *CAA Annual Statistics* [QRL.20] and [QRL.19].

3.2.5 *Travel*

3.2.5.1 General. The *UK Balance of Payments* 'Pink Book' [QRL.61] examines Travel Services on an annual basis, giving credits, debits and balances by purpose of visit; business, leisure and total. The figures cover goods and services provided to UK residents during trips of less than one year in overseas countries and provided to overseas residents during similar trips in the UK. The figures are net of any purchases met out of money of a local source. The 'Pink Book' figures exclude transport to and from the UK; where these affect the UK balance of payments, they are included in 'sea transport' and 'civil aviation' (see sections 3.2.3 and 3.2.4 above). The figures do include transport services provided overseas to UK residents that are not already included in the price of the ticket for the journey to or from the UK and similar transport services provided to foreign residents within the UK. These estimates are based primarily on the International Passenger Survey (IPS), (see section 3.2.5.2 below).

The IPS survey does not include travel between the UK and Irish Republic. Estimates of this expenditure are derived from statistics published by the Irish Central Statistics Office. In the 'Pink Book' 'leisure' includes all purposes of a visit except 'business'. All overseas visitors staying overnight in the UK *en route* to other destinations are included under leisure. Any more detailed study of travel needs to refer to the data presented in more detail in the sources analysed in the following sections.

3.2.5.2 The International Passenger Survey (IPS). The International Passenger Survey (IPS) is carried out for the Department of Employment, the Central Statistical Office and the Home Office by the Office of Population Censuses and Surveys and analysed by the Department of Employment. It is not separately published but the main results are presented in *Business Monitor Overseas Travel and Tourism MA6* [QRL.11] on an annual basis and a more detailed breakdown of quarterly figures is given in *Business Monitor MQ6* [QRL.13].

The estimates are based on interviews with a stratified random sample of passengers entering and leaving the UK on the principal air and sea routes. Interviews are voluntary and anonymous. The main features of stratification relate to transport mode (air or sea), port, and time of day. The frequency of sampling within each stratum depends, mainly, on the variation of tourist expenditure and on the volume of migrants (see Chapter 10). Travellers passing through passport control are randomly selected for interview (169,000 interviews in 1985). Only interviews at the *end* of a visit provide information on expenditure and length of stay.

These results are supplemented with estimates from the Central Statistics Office of the Irish Republic of travel between the UK and the Irish Republic. Figures for the Channel Islands are provided by the Economic Advisory Office of the States of Jersey.

Approximately 90 per cent of passengers entering or leaving the UK (excluding journeys to and from the Irish Republic) travel on routes covered by the survey. The remainder are either night travellers or on routes with too small a volume to be covered. Estimates are made in these cases. Percentage coverage of travellers is given in [QRL.11]; in the 1985 survey it was 0.24 per cent. A complex weighting procedure is used in the survey results taking account of passenger movement statistics produced by the Civil Aviation Authority (air traffic) and Department of Transport (sea traffic). Allowances are made for passengers in transit at Heathrow and Gatwick Airports.

Accuracy of the results from the IPS. Collection of the information by sample survey means that estimates of number of visits, length of stay and expenditure are subject to a sampling error. This error varies according to the mix of ports used by a particular category of traveller because the intensity of sampling varies at each port. In general, the major airports and long-haul sea routes are most frequently sampled and these estimates are most accurate. Other influences create errors such as response errors, design of the sample and weighting and grossing-up procedures. These cannot be quantified and so the errors quoted (see [QRL.11] Introduction) are minimum errors.

Sampling errors of the IPS are given in terms of an interval into which there are 19 chances out of 20 that the true figure will fall. The estimate plus or minus the percentage quoted gives the appropriate interval. For 1985 as a whole, the estimates of the total number of overseas residents visiting the UK (all figures exclude the Irish Republic) are judged to be correct within an interval of plus or minus 1.1 per cent. The sampling errors vary widely e.g the number of visitors by country of residence varied from plus or minus 2.1 per cent (USA) to plus or minus 28.8 per cent (Yugoslavia). The estimate for total UK residents going abroad has a sampling error of plus or minus 1.0 per cent for the whole of 1985. Quarterly estimates for all figures, of course, have a wider margin of error. Detail on the sampling errors of the IPS is given in *British Business* 17th August 1984 [B.30].

3.2.5.3 Definitions. Numbers in [QRL.11], [QRL.13] are numbers of visits not visitors; anyone entering or leaving more than once in the same period is counted on the occasion of each visit.

Day trips not involving an overnight stay are included in the figures for visits and expenditure. Details of such visits are shown separately in some tables as 'excursionists'. This does not cover day-trips to or from the Irish Republic although these trips are included in total visits. Overseas residents in transit in the UK but who do not stay overnight are excluded but spending whilst in the UK is included in the figures for earnings.

Visits are for a period of less than 12 months. When a UK resident visits more than one country, the *entire* visit, expenditure and length of stay are allocated to that country in which the resident spent the longest time.

3.2.5.4 Travel and Tourism: Annual Figures. Business Monitor MA6 *Overseas Travel and Tourism* gives annual figures on travel and tourism derived from the International Passenger Survey (IPS) [QRL.11]. The figures are divided into two main sections: (a) overseas visitors to the UK and (b) visits abroad by residents of the UK. A summary table gives basic comparative figures for an eighteen-year period. In all cases the latest year's figures are provisional and liable to revision.

(a) *Overseas visitors to the UK.* The data are given by number of visits and expenditure for the past three years. Number of visits are broken down for the most recent year by air and sea and country of permanent residence (and area). The average length of stay in days, average expenditure per visit and average expenditure per day is given for the most recent three years. Average length of stay, expenditure per visit and per day by country of permanent residence is given as is number of visits by area of permanent residence, main purpose of visit and mode of travel (all routes, air, sea).

The main purpose of the visit is classified into: holiday visits, business visits, visits to friends and relatives (VFR) and miscellaneous visits. Visits for miscellaneous purposes include those for study, to attend sporting events, for shopping, health, religion or other purpose, together with visits for more than one purpose when none predominates. Overseas visitors in the UK *en route* to other destinations are also included in miscellaneous purposes. It is further to be noted that estimates relating to travel flows across the land boundary between Northern Ireland and the Irish Republic are 'for convenience' included in the figures for sea. Hovercraft flights are also classified as sea crossings. Holiday visits have 'inclusive tours' separated out, which makes it possible to separately classify and calculate the importance of this type of holiday.

Further cross-classifications give (i) average length of stay, expenditure per visit and per day by permanent residence and main purpose of visit and (ii) number of visits and expenditure by quarter, area of permanent residence and main purpose of visit. Number of visits by country of residence is also classified by main purpose of visit (all visits, holiday visits, inclusive tour holiday visits, business visits, VFR and miscellaneous visits) for the most recently available three years.

The number of overnight visits to the regions of the UK by country of residence are given — separately for London, other England, Scotland and Wales. These exclude visitors from the Irish Republic and all visits not including an overnight stay in the UK. Although included in the 'total' column, visits to Northern Ireland are not separately analysed. As more than one region can be visited by an individual whilst in the UK, the total of the visits to all the regions is greater than the total number of visits to the UK.

(b) *Visits abroad by residents of the UK.* The data are classified similarly to that for overseas visitors *to* the UK. Number of visits and expenditure abroad by main country visited and mode of travel are given for the latest three years as are average length of stay, expenditure per visit and per day by main country visited. Number of visits by area visited, main purpose of visit and mode of travel are separately classified as are average length of stay, expenditure per visit and per day by area visited and main purpose of visit. Quarterly classification for the most recent year are given for number of visits and expenditure by area visited and main purpose of visit. Number of visits by country and main purpose of visit are given for the past three years with holiday inclusive tour visits separately identified.

3.2.5.5 Travel and Tourism: Quarterly Figures. Visits to the UK by overseas residents and visits abroad by UK residents are given quarterly by area of residence and are listed together with the percentage change in total on the previous year [QRL.13], see also [QRL.7]. Overseas earnings and expenditure are provided both in current and at constant (1980) prices adjusted and unadjusted. Visitor nights in UK/abroad by overseas visitors/UK residents are also given on a quarterly basis [QRL.13] and [QRL.7].

The number of visits to the UK by overseas residents by all modes of travel is broken down by main purpose of visit into four chief categories: holiday, business, visits to friends and relatives (VFR) and miscellaneous. Excursionists and inclusive tour holidaymakers are separate sub-categories. These quarterly figures are also subdivided by area of residence which are: North America, EC, other Western Europe, other areas. These figures are further available according to mode of travel: sea and air [QRL.13] also [QRL.7]. Further information on the origin of visitors to the UK by country of residence is cross-classified with mode of travel.

The destination of overseas visitors (defined to include excursionists and tourists) is given by the number of overnight visits to the regions of the UK, by area of residence.

The number of visits abroad by UK residents by mode of travel (sea, air) and by main purpose of visit and area visited is available on a quarterly basis [QRL.13] and also [QRL.7]. The former publication allows a breakdown by country visited on a quarterly basis and by country visited cross classified by mode of travel (air, sea) on an annual basis.

Annual figures on travel and tourism [QRL.11] give a more detailed analysis of expenditure than the quarterly ones [QRL.13].

3.2.5.6 Travel and Tourism: Monthly Figures. Overseas travel and tourism details are available on a monthly basis, covering the number of overseas visitors to the UK and visits abroad by UK residents, together with earnings and expenditure (total earnings, expenditure and balance) [QRL.13].

Detailed breakdowns of overseas travel and tourism are given in the Department of Employment's *Press Notice on Overseas Travel and Tourism* [QRL.50]. Initial estimates of overseas visitors to the UK and visits abroad by UK residents are given on a monthly basis in [QRL.7] by area of residence or area visited. Similarly, initial estimates of overseas earnings from and expenditure on overseas travel and tourism are given in total with the balance. Monthly figures on overseas travel and tourism earnings and expenditure at current prices, numbers of visits to UK and visits abroad are given in [QRL.32]. Quarterly figures on the purpose of visits are given in less detail also.

3.2.5.7 Travel and Tourism: European Community. The Working Group of the National Tourist Organisations of the EEC produced a Report *The Economic Significance of Tourism within the European Community* [QRL.29]. Details of its contents are listed in Appendix 3.

3.2.5.8 Regional Information. The regional data referred to in this section are on a very broad brush basis: London, 'other England', Scotland and Wales. However, each Regional Tourist Board publish 'Tourism Fact Sheets' giving a regional breakdown on an annual basis. These are published by the British Tourist Authority, English Tourist Board Research Services.

Tourism Fact Sheets [QRL.59] have been produced for each of the English tourist regions since 1975. The principal sources of information are the British Tourism Survey and the International Passenger Survey. The British Tourism Survey is conducted annually and together with a regular monthly survey based on personal interviews, in published annually as *The British Home Tourist Market* [QRL9]. Summary findings are published in the *Digest of Tourism Statistics*. The International Passenger Survey [QRL.11] is described in 3.2.5.2 above.

3.2.6 *Financial and Other Services*

3.2.6.1 UK Financial and Allied Institutions. A detailed annual breakdown of services, financial and other, is given in the *UK Balance of Payments* [QRL.61]. Earnings are given by type of institution, firm or service. Net credits are subdivided by UK Financial and Allied Institutions covering insurance underwriters, insurance brokers, banks, commodity traders, export houses, other brokers etc. and the total is given.

This information derives from a variety of sources, and these sources and the nature of the estimates must be assessed for likely accuracy.

Underwriters' services are measured by net premiums from overseas (premiums less claims paid) less overseas expenses. The figures for Lloyd's underwriters are supplied by the Corporation of Lloyd's Insurance Companies. Data is supplied by the Association of British Insurers and the most recent year quoted may be a projection. *Brokers' services* are measured by net brokerage earnings on business written in foreign currencies and sterling business known to relate to overseas residents. These figures are estimates based on returns to the Department of Trade and Industry.

UK banks' services are based on a periodic enquiry carried out by the Bank of England. Following the latest survey, estimates for earlier years are reassessed. Banks' management services are included under 'other services' provided by/to UK companies to/by related concerns.

Brokerage services by UK firms, other than oil companies, in respect of physical trade in commodities (industrial raw materials and foodstuffs) traded between foreign countries and transactions on commodity futures markets are covered. The estimates are provided by the Bank of England and for 1974-9 are based primarily on a survey of the principal commodity markets which the bank carried out for three years from mid-1974 (some results were given in [B.27]). Not all commodities were covered by the survey and an allowance is made based on broad indicators from some of the trade associations concerned. From 1980 onwards information on total turnover (in value terms) on the London commodity futures markets is used to update the estimates. These figures can thus only be regarded as giving broad 'ballpark' figures.

Estimates of the profits of *export houses* are based on occasional surveys carried out by the British Export Houses Association. The latest survey used by the 1987 'Pink Book' [QRL.61] is in respect of 1977. The estimate has simply been projected forward based on the trends in the value of world trade. Again, because of the age of the latest basic estimate, extreme caution must be used in respect of the estimate.

Other brokers estimates are supplied by the Bank of England in respect of the Baltic Exchange, the Stock Exchange, Lloyd's Register of Shipping and Other Brokers. Returns by broker members of the Stock Exchange ceased after the first quarter of 1980, pending the introduction of a new reporting system, covering all security deals, following the deregulation of the Stock Exchange ('Big Bang') and projections have replaced direct reporting.

3.2.6.2 Other Services. UK *consultancy firms* [QRL.61] gives net credits for UK consultancy firms: consulting engineers, process engineers, solicitors and barristers, management and economic consultants, chartered surveyors and other not elsewhere identified (NEI) together with the total.

Annual estimates of the net earnings of consulting engineers, chartered surveyors and management and economic consultants are supplied by professional associations. Estimates of the gross earnings of process engineers are provided by industrial firms. Estimates up to 1977 of the earnings of solicitors are based on VAT returns — later returns are based on surveys held in 1981 and 1987 (in which amounts forwarded to barristers were included). The other identified (NEI) category contains annual estimates provided by the Nationalised Industries' Overseas Group for those of its current membership whose earnings are not included elsewhere, estimates of the earnings of architects and a national account, based on exchange control records covering miscellaneous professional earnings. The arbitrary nature of many of these figures can easily be seen.

Telecommunications and postal services figures are supplied by British Telecom and the Post Office.

Films and television figures in the 'Pink Book' [QRL.61] are derived from the annual Department of Trade and Industry inquiry into overseas transactions in respect of production costs, rentals and purchases of cinematograph films and television material as covered in detail in [QRL.10]. A deduction is made to eliminate the overlap between this inquiry and the purchases and sales of exposed cinematograph films included in exports and imports of goods. Full details are given in section 6.3 below.

The main credit items of *North Sea Oil and Gas Companies* are work done overseas by drilling rigs, supply boats etc. The main debit item, here, is services such as of drilling rigs, vessels, consultancy, diving, etc. The basic information is supplied by companies classified to this industry (section 13 of SIC (1980)).

Credits of *Land Transport-Freight* are the earnings of UK road hauliers for carriage outside the UK of UK exports and for carrying cross-trades. An entry is also included for their earnings from carrying UK imports within the exporting country. Estimates of tonne-kilometres are derived from the International Road Haulage Survey and freight rates are 'estimated from trade sources'.

Freight rates are estimated from trade sources. Debits include payment to all overseas land transport operators for the carriage of UK imports between the exporting countries and the UK. Estimates are made by subtracting from the estimate of total freight on import from land less an estimate of the element provided by UK operators (this has changed since 1982, before then some of these payments were included under 'sea transport'). Debits also include an estimate of overseas road hauliers for carrying UK exports and imports within the UK. Carriage across the land border in Ireland is excluded. Spending by UK road hauliers abroad and foreign hauliers in the UK is included in 'travel'.

The credit representing *Advertising* services provided for non-residents by the UK media and by advertising agencies is based on information forwarded by Media Expenditure Analysis Limited.

Overseas *Royalties and other Services* reported in the UK *Balance of Payments* [QRL.61] are based on an annual voluntary enquiry held by the Department of Trade and Industry, supplemented with data from specialist computer firms, auction houses and individual companies. These are projected forward from the previous year's survey using trends in industrial production, overseas prices and exchange rates. These figures are thus estimates for the current year. Where exchange rates, for instance, are volatile there will be inaccuracies.

Separate figures are shown for transactions with related (overseas parent companies, branches, subsidiaries and associates 'the definition widened in 1979') and unrelated concerns. A full analysis of royalties is given in [QRL.10] and chapter 6 below. Banks, oil companies and insurance companies are excluded. Royalties cover amounts payable or recoverable for the right to use processes and other information. Film royalties are included under 'films and television' and royalties incorporated in the contract prices of UK exports and imports are included in visible trade.

Other sources are those not covered elsewhere in section 3 above (invisibles: services). Services to and from related concerns cover both charges for general management expenses and for specific services. They exclude UK oil companies and services shown elsewhere. Unrelated concerns include *all* royalty and services transactions of UK oil companies, even those with related concerns. Estimates of services provided by unrelated concerns are included with miscellaneous debits. Payments for royalties and other services received or made locally are, like other trading transactions, included under interest, profits and dividends (section 5 of the Balance of Payments).

Fees and other expenditure in the UK by overseas students, overseas embassies etc. US bases and forces (NEI) are obtained by applying estimates to the numbers of overseas students. Estimates are provided by the British Council and various educational boards. Expenditure by overseas students for a period of stay shorter than one year is included in 'travel'. Estimates of expenditure by overseas embassies are based on information provided by certain overseas countries on numbers of diplomats and UK prices based on a sporadic survey. US military authorities supply information on expenditure in the UK.

Commission on UK imports: estimates are made of agent's services to suppliers and after sales services by applying 'appropriate ' percentages to c.i.f. values of various commodities derived from sample surveys. The last of these (as of 1987) relates to 1980.

Miscellaneous credits include the earnings of foreign workers working temporarily overseas, subscriptions to UK professional associations and other minor receipts. Debits comprise an estimate of the value of services provided to UK companies by unrelated overseas concerns and other miscellaneous payments. The estimates are based on exchange control records up to 1979, updated from various sources, including the DTI annual survey. Like the above smaller elements in service earnings, they are unlikely to be accurate because of the lack of direct data gathering and the use of leasing related series to update the basic, elderly estimates.

Loss/damage to UK imports includes insurance services on UK imports provided by overseas insurers.

3.3 Invisibles: Earnings on UK External Assets and Liabilities (Interests, Profits and Dividends)

Earnings from the ownership of foreign assets are in the form of interest, profits and dividends. The *United Kingdom Balance of Payments* the 'Pink Book' [QRL.61] distinguishes two major groups: earnings from direct investment and from portfolio investment together with a category of other earnings on external assets.

The flow of investment is separate from the earnings, although unremitted profits are a category of both. Investment, as a capital term in the balance of payments is covered in chapters 4 and 5. Earnings are defined to include all profits earned and interest and dividends due, whether paid or not. They are recorded after deduction of local taxes, and, for profits, after allowing for depreciation.

Profits and dividends include the (credit) earnings from overseas affiliates of UK registered companies and the (debit) earnings of profits and dividends by UK based affiliates of foreign based companies. As the 'Pink Book' [QRL.61] points out, conceptually, stock appreciation and other unrealised capital gains and losses should be excluded from the flows in the balance of payments accounts, because they represent changes in asset valuation. In practice, these will be included in earnings and offset as reinvested profit/losses in the capital account. Profits retained abroad by overseas affiliates or retained in the UK by foreign owned companies are included in the flows of earnings and offset in the capital account.

The *UK Balance of Payments* [QRL.61] gives a summary of earnings on UK external assets and liabilities (interest, profits and dividends, (IPD)) on an annual basis. This covers direct investment, portfolio investment and other earnings by type of asset or liability. It includes total and net earnings on assets/liabilities of: general government, public corporations, UK private sector dividing this by assets by banks/non-bank.

Interest, profits and dividends, credits, unadjusted and seasonally adjusted, are given on a quarterly basis [QRL.30]. These credits are earnings on direct investment (non-oil companies, oil companies and miscellaneous, total), portfolio investment, export credit, external foreign currency lending by UK banks, external sterling lending by UK banks, bank deposits and lending overseas, official reserves, other assets and total credits (also divided into private sector and public corporations, general government). These data are similar to [QRL.62] but include an unadjusted figure, oil/non-oil split and official reserves.

The debits and balance, for IPD unadjusted and seasonally adjusted are similarly given in [QRL.30]. Debits are for direct investment (as above), portfolio investment, external foreign currency borrowing by UK banks, external sterling liabilities, borrowing from banks etc. overseas (but excluding loans to HMG and under the public sector exchange scheme which form part of official funding), official financing liabilities (excluded from the otherwise similar data in [QRL.62]), other liabilities.

Total debits are divided as in the paragraph above and the balance of IPD is given for private sector and public corporations, general government and total.

3.3.1 *Earnings on Direct Investment*
Earnings on direct investment are given separately in [QRL.61]. Earnings on investment overseas by UK residents is given by companies and public corporations subdivided into: unremitted profits, dividends remitted, net interest received and profits of branches. This includes the direct investors' portion of reinvested earnings, which also appears in the capital account as an offsetting entry (see chapter 4).

The companies' own estimates for depreciation are used and although depreciation is estimated at replacement costs in the UK National Accounts [QRL.64], in the main, these are entered in the balance of payments at historic costs. Depreciation is measured after provision for local taxes. Refunds of tax retrospectively made under double taxation agreements are included in the period when they were made rather than the earlier accrual period. Figures on refunds to overseas investors are supplied by the Inland Revenue.

Both credit and debit earnings of all companies except banks and insurance are obtained from the Department of Trade and Industry's Direct Investment Voluntary Inquiry. Estimates are made to cover non-reponse. Each quarter a small sample of companies submit returns providing quarterly estimates which become available about two months after the end of the quarter. The estimates for the latest year are based on the results of these quarterly enquiries. Results of the annual enquiry are available from about eighteen months after the end of the year and are published in *British Business* [QRL.7] and *Business Monitor MA4* [QRL.10] (see also sections 5.3 and 5.4 below).

In 1985 the DTI inquiry was extended to cover the oil companies' overseas transactions. Oil companies were surveyed retrospectively for their 1984 transactions and these are shown. Figures for oil companies' earnings from 1984 are published in [QRL.7].

3.3.1.1 IPD — Oil Companies. Prior to the DTI inquiry covering oil companies (1985) and the retrospective survey covering 1984, UK oil companies' earnings were included in 'earnings on overseas assets by other UK residents'.

Treatment of oil companies before 1984 is the subject of a lengthy note in the 1987 'Pink Book' (page 30) [QRL.61]. A discontinuity in the series [QRL.61] tables these results.

3.3.1.2 Banks and Insurance Companies. Information on the direct investment earnings of UK registered banks from their overseas registered banks, from their overseas branches, subsidiaries and associates and of overseas registered companies from their UK banking affiliates is collected in an annual inquiry carried out by the Bank of England. An annual inquiry by the Association of British Insurers (ABI) on behalf of the DTI forms the basis of estimates for direct investment earnings by UK companies. This inquiry was extended in 1984 and previous years' figures incorporate estimates to allow for change in coverage. These elements of IPD in [QRL.61] can therefore be regarded as up-to-date and accurate, as they are based on direct inquiries.

3.3.1.3 Earnings and overseas assets by other UK residents and earnings on UK liabilities: miscellaneous property investments in the UK. Earnings on overseas property by UK resident financial institutions and persons are estimated from the levels of such assets held by financial institutions and information on the total overseas income of the institutions. It should be noted that the property earnings of non-financial institutions are included under 'companies' indistinguishably. The overseas earnings of construction contractors included here are an estimate of part of their direct investment income which was covered before 1987 within the category of service earnings. This figure is included with industry and commercial companies in the sector analysis of direct investment but is *not* included in the table of earnings on direct investment in the 1987 'Pink Book' [QRL.61].

Miscellaneous property investment earnings in the UK comprises 'notional estimates' of interest due to overseas owners of residential property in the UK [QRL.61].

3.3.2 *Earnings on Portfolio Investment*

Portfolio investment occurs when the individual, company or institution making the investment does not control the investment. Control is the defining characteristic of direct investment.

Earnings on portfolio investment are given by type of security in [QRL.61]. Credits are earnings (interest and dividends) on overseas government and company securities by earning sector, viz: UK banks, other UK financial institutions, other UK residents. Debits are earnings (interest and dividends) on overseas investment in the UK. These are subdivided by type of asset: British government stock (earnings by overseas authorities, other overseas residents), British government foreign currency bonds and notes, local authorities' securities (sterling, foreign currency under the exchange cover scheme, other foreign currency), public corporations' securities, (issued under the exchange cover scheme, other (foreign currency)) UK companies' securities (issued under the exchange cover scheme, other bonds etc, ordinary shares) and total.

These figures are also classified as total earnings on securities issued by: general government, public corporation, UK companies [QRL.61].

A large part of the total credits is received by UK financial institutions not involving control and is classified as portfolio rather than direct investment. These figures are largely estimated from statistical inquiries of banks and other financial institutions together with estimates of asset levels and rates of return for other UK residents. In early years, information from the Inland Revenue was also used.

3.3.3 *Other Earnings on External Assets of the UK and Other Earnings on UK Liabilities to Overseas Residents*

3.3.3.1 Other Earnings on External Assets of the UK. This credit term [QRL.61] includes: lending to overseas residents by UK banks, deposits and lending overseas by UK residents other than banks and general government (transactions with banks abroad and other assets), official measures and other external assets of central government [QRL.61].

UK bank lending to overseas residents: the credit here is interest and it is based on returns provided by the Bank of England. Changes in this series includes the following: (i) interest on bank lending in foreign currencies, other than long-term credit, is included on a gross base for 1972 onwards only in the summary table of the current account (Table 1.3 of the 1987 'Pink Book' [QRL.61]). In earlier years it is deducted from the corresponding debit item, (ii) miscellaneous financial institutions' earnings are also included up to the end of 1981, from 1982 they are included with those of other UK residents.

Earnings from banks abroad are included in 'interest on deposits and lending overseas by UK residents other than banks and general government' from the beginning of 1976. From 1982 estimates of the appropriate earnings of miscellaneous financial institutions, previously included indistinguishably with UK banks, as above, are included in this category.

Earnings on other non-bank private sector and public corporations assets are estimated from rather dated information from the Inland Revenue, Exchange Control information and specialist sources. They are likely to be estimated in future from rather ageing and disparate basic information and should be used with great care.

Earnings on official foreign exchange resources and on the UK's holding of Special Drawing Rights (SDRs) is recorded within the Exchange Equalisation Account by the Bank of England. Earnings on other external assets of central government are reported by the Government Department responsible for the loans. Refinancing by ECGD is included.

3.3.3.2 Other Earnings on UK Liabilities to Overseas Residents This debit item includes: borrowing from overseas residents by UK banks, borrowing from overseas by UK residents other than banks and general government (transactions with banks abroad, other liabilities) and other external liabilities of general government [QRL.61].

Interest on borrowing from overseas residents by UK banks is estimated from banking statistics and the series dates back only to 1983; prior to this, interest paid by miscellaneous financial institutions was included.

Interest on overseas borrowing by UK residents other than banks and general government is interest payable to banks plus that payable to non-banks. Generally, it is estimated by applying an appropriate interest rate to estimates of levels outstanding.

Interest on other external liabilities for general government includes that on IMF loans, expenses for use of SDRs, long term fixed interest loans (e.g lend-lease), interest on Treasury bills and local authorities' banks. Figures are available for these from the Bank of England and the Treasury and supplied for balance of payments purposes.

Earnings on liabilities of general government (central and local) are a separate category. Debits cover interest paid on overseas holdings of British government stocks, British government foreign currency bonds and notes, Treasury bills, and other UK central and local assets held abroad.

3.3.4 *Summary*

IPD, earnings of UK external assets and liabilities are classified into the total earnings on assets, on liabilities and net earnings of (i) general government (ii) public corporations and (iii) UK private sector in the 'Pink Book' [QRL.61]. This covers direct investment earnings, portfolio investment earnings and other earnings.

3.4 The Net Overseas Earnings of UK Financial Institutions (the City)

A separate table in the 'Pink Book' brings together all the identifiable current account transactions of City institutions (Table 6.1 in the 1987 edition [QRL.61]).

This table cuts across two types of invisible transaction: services and receipts from investments (IPD). It thus re-interprets entries included in sections 3.2 and 3.3 above. The City is interpreted by nature of institution and not by geographical location. The institutions covered are insurance, banking, leasing, commodity trading, investment trust companies, brokers and security dealers.

The meaning of net earnings is important enough to deserve a separate note in the 'Pink Book' [QRL.61]. For financial institutions, service earnings are entered only after deduction of relevant overseas expenses have been deducted. The argument is that this provides an estimate of the value of the service component. Service earnings are thus recorded 'net'. The debit entries in The City table are the investment income due abroad from the UK insurance and banking sectors and the interest payments on borrowing from non-residents by UK banks and these are deducted to arrive at the 'net' earnings of financial institutions.

3.4.1 *Coverage*

As the table on 'the City' is a compilation of items from elsewhere in the balance of payments the description here can be brief.

The areas covered are insurance, banking, leasing, commodity trading etc., investment trust companies etc., brokerage etc. and securities dealers.

Insurance earnings cover underwriting and brokerage earnings plus direct investment, earnings and property income. Portfolio investment is also covered with information from new insurance inquiries. Lloyd's (which is included in the personal sector in the IPD section of the 'Pink Book' [QRL.61]) is also covered.

Banking figures are the net receipts from the overseas operations of all banks in the UK including those of branches in the UK of foreign-owned banks. To these receipts are added the total profits of British-owned banks overseas *less* the total profits earned by UK branches and subsidiaries, associates of foreign owned banks, whether or not these arise from international or domestic transactions. The treatment varies from the IPD section of the 'Pink Book' as it is more narrowly defined. Information is essentially derived from the Bank of England but it was also based on a DTI annual enquiry which ceased at the end of 1982.

Leasing covers only the earnings of specialist finance leasing companies. In the UK balance of payments accounts leasing is treated as if it were lending by the leasing company to the leasee to finance his purchase of the asset. Estimates of the interest received by leasing companies on their loans are included indistinguishably with the capital repayments.

Commodity trading earnings are estimated by the Bank of England and include profit on third country trade, consumer earnings on futures markets and earnings on 'differences'.

Investment trust companies make annual returns and local authority pension funds are estimated 'using information on overseas holdings from annual balance sheet enquiries' (estimates only before 1982). All overseas income is included less estimated interest paid on foreign currency borrowing. Payments abroad in respect of dividends, income distributions, pensions, etc. are not included but are believed to be small.

Brokerage earnings are net overseas receipts of the Baltic Exchange, Stock Exchange, Lloyd's Register of Shipping and other brokers, which includes gold and silver bullion dealers, and the net overseas receipts of foreign exchange brokers.

Securities dealers' earnings include portfolio earnings. Securities dealers are only classified as financial institutions from the beginning of 1984. New reporting regulations in the Stock Exchange are to take effect in the near future to the figures under Stock Exchange Brokers are only estimates.

3.5 Transfers

Transfers are unusual items in the balance of payments accounts. Most entries represent resources provided and the payments for those resources or changes in financial assets and liabilities. Generally, therefore, two entries occur in the accounts. For some transactions, only one entry appears — a resource provided or a change in financial assets. Examples are gifts of goods to overseas residents (a positive entry under visible trade or services) and simple transfers to overseas residents of financial assets (a positive entry in the capital accounts). The entries under transfers in the balance of payments represent the counterpart items to these entries (the value of the gifts of goods or of the assets, negatively signed, in the examples). Also included are the counterpart terms to the capital account entries resulting from money being brought to, or taken from, the UK by migrants. (No entry is made for 'temporary migrants' assets retained in the home country.)

All transfers are included in the current account. Those of a capital nature are, in practice, difficult to distinguish as such [QRL.61].

3.5.1 *General Government Transfers*

General government transfers cover grants to overseas countries, subscriptions and contributions to international organisations, other transfer payments to overseas residents and similar receipts from overseas by the UK government. Credits are from European Community institutions. Debits cover contributions and subscriptions to international organisations broken down into: EC institutions, military, multilateral economic assistance and 'other' categories. Also covered are bilateral aid: project grants, non-project grants and technical cooperation plus military grants, social security benefits and other transfers. These categories plus total government debits and the general government transfer balance are tabulated from official records [QRL.61]. They can therefore been assumed to be accurate.

3.5.2 *Private Transfers*

Private transfers, of necessity, rely on 'guesstimates' to a large degree [QRL.61]. Credits comprise: savings from work of UK nationals in the Middle East (numbers multiplied by assumed savings at a 'recent date' trended forward by the global transfer of debits of the host countries), pension payments and transfers from OECD countries based on information supplied to the UK and, transfers from other countries, which are estimated by applying appropriate percentages to debit figures of nine countries with a percentage addition for the rest of the world, and assuming an average amount of assets per migrant. Debits include: cash gifts from UK households to dependants abroad and legacies (from Exchange Control records to 1979, extrapolated by Family Expenditure Survey Data and number of deaths), payments abroad by voluntary aid agencies (data supplied by the agencies), emigrants' assets (based on Exchange Control records and International Passenger Survey data), the gift element in the parcel post (based on a sample enquiry held by the DTI) and pensions paid abroad (estimated by fund managers). The nature of this information clearly conveys the message that use of this data and its interpretation requires great caution. Its quality like many other international sources, has clearly deteriorated since the end of Exchange Control (1979).

3.5.3 *Other Information on Transfers*

A quarterly breakdown of transfers, seasonally adjusted and unadjusted is provided in *Economic Trends* [QRL.30]. Credits are broken down by category (i) private, (ii) general government covering EC institutions: agricultural fund, social fund, regional development fund, budget refunds, other and total. Debits are (i) private, (ii) general government: bilateral aid, EC institutions, other, total. The balance is given for both private and general government; also the total balance of which that with EC institutions is shown separately [QRL.30]. A very similar breakdown is provided in [QRL.62] but it exludes unadjusted figures and gives a less detailed breakdown of receipts from EC countries.

General goverment transfers on an annual basis are given in *Consolidated Fund and National Loans Fund Accounts* [QRL.23] and *Consolidated Fund and National Loans Fund Accounts Supplementary Statements* [QRL.24] which cover subscriptions and contributions to international organisations. The total of payments from the Consolidated Fund to the EC etc., given in [QRL.23], is subdivided into Budget of the EC and European Investment Bank Capital Subscription in [QRL.24]. This source also gives a further breakdown of the Budget of the EC with agricultural levies, customs duties and VAT contributions, in particular, being shown individually.

3.6 Invisibles: Final Comments

The make up of invisible trade and its constantly changing nature give rise to large problems in (i) definition and (ii) measurement. Table 1.2 in chapter 1 shows a (dated) estimate of the likely reliability of information and this reliability has declined in many areas with the ending of Exchange Control regulations and the consequent abandonment of recording. Caution is particularly required with regard to information on 'other services' transfers, and particularly private transfers, and therefore on the current balance. Appendix 1 gives an indication of the nature of projections in these areas: 40 per cent of services credits are based on data which are reported either annually or less frequently, or sometimes not reported at all. This figure is 44 per cent for transfers (all private transfers), 10 per cent for IPD, 28 per cent for service debits, 35 per cent for transfer debits and 7 per cent for IPD debits. Definitional problems, leading to coverage difficulties, arise particularly in areas where new financial instruments are being developed.

The presentation of figures on a net basis e.g for the 'City's' foreign earnings has been criticised. This is rather more an issue of the *use* of statistical information than its presentation. In the area of international comparisons, however, it is essential to compare like with like so gross and net figures should be separately tabulated and separately compared. International comparisons in the invisible exports field are even more than usually fraught with difficulties of definition, coverage, differential accuracy of collection and of quality of basic data.

Other useful information on invisibles comes from the British Invisible Exports Council, notably *World Invisible Trade* [QRL.67] and occasional special reports e.g *International Information Flows* (report by the Liberalisation of Trade in Services Committee (LOTIS)) [B.41]. An analysis of exports and imports of services by industry has also been carried out [B.21] and the *UK National Study of Trade in Services* [B.44] provides information on each sector plus an outline of the organisational structure and major trading partners involved. Annexe A gives an outline of the principal statutes and regulations governing service sectors in the UK and the powers of UK regulatory bodies in this area. A new survey of invisible trade is also available [B.13].

CHAPTER 4

INTERNATIONAL CAPITAL MOVEMENTS

4.1 Introduction

This chapter serves as an introduction and overview to Chapters 5, 6, 7 and 9 which concern the details of capital movements to and from the United Kingdom.

The *UK Balance of Payments* [QRL.61] is presented in two main sections: current account transactions (considered earlier in Chapters 1, 2 and 3) and transactions in UK external assets and liabilities. This latter category covers inward and outward investment, not: external transactions by banks in the United Kingdom; borrowing and lending overseas by other UK residents and drawings on, and accruals to, the official reserves and other capital movements.

4.2 Transactions in UK External Assets and Liabilities

The presentation of transactions in UK external assets and liabilities in the 'Pink Book' [QRL.61] follows the broad outline for the capital account of the balance of payments as set out in the IMF *Balance of Payments Manual* [B.26]. Transactions are identified under: (i) direct investment, (ii) portfolio investment, (iii) other capital excluding (official) reserves and (iv) (official) reserves. Summary tabular presentations by UK sector are given separately for: direct investment (together with a sectoral analysis), portfolio investment, transactions of UK banks, transactions of UK residents other than banks and general government [QRL.61], (see also section 1.5 above for a description of the breakdown). A table of overseas authorities' identified exchange reserves in sterling is also given.

The term investment has a wide meaning in the context of balance of payments accounts. It refers not only to the creation of physical assets but also to the purchase and sale of paper assets such as stocks and shares and other securities. Direct investment also covers the financing of trade movements and other financial transactions between related companies in the UK and abroad.

Other capital transactions consist mainly of borrowing or lending by banks which may be associated with UK visible trade. Borrowing or use of deposits will be included in the appropriate term in the capital account offsetting the entry under visible trade. Banking transactions similarly may arise from the financing of other capital transactions. These factors serve to emphasise the interdependencies within the balance of payments accounts which were emphasised, together with the problem this raises, in chapter 1 (section 1.1.4).

Total (stock) values of assets and liabilities at year end are covered below in section 4.3. Income earned from them is covered in chapter 3 (section 3.3). Chapter 5 covers annual transactions and book values of inward and outward direct foreign investment. Other overseas transactions and international transactions of the money and banking sector are considered in chapters 6 and 7 respectively. The remainder of this section examines balance of payments information.

4.2.1 *Direct Investment*

The balance of payments 'Pink Book' [QRL.61] gives investment overseas by UK residents for (1) by companies and public corporations and (2) by other UK residents. Oil companies were included in the second category before 1984 but are now in the first category. Separate figures for the oil companies' investment from 1984 are published in *British Business* [QRL.7]. The total for companies and public corporations is broken down into: unremitted profits of subsidiaries, net acquisition of share and loan capital, change in inter-company accounts and change in branch indebtedness. The second category is undivided.

Investment in the UK is divided into: in non-oil companies (broken down as for companies and public corporations above), in oil companies (unremitted profit of subsidiaries, other investment and total) and miscellaneous property investments.

A sectoral analysis of overseas investment by UK residents provides the following breakdown: UK banks, other financial institutions, industrial and commercial companies (including oil companies) and persons, public corporations and total. Investment in the UK is broken down into: UK banks, other financial institutions, industrial and commercial companies (including oil companies) and other (property) [QRL.61].

In 1984 new enquiries concerning direct investment by insurance companies were introduced.

It should be noted that when the DTI's financial survey into external liabilities becomes available direct investment flows are revised back to the date of the last survey. This survey is now not quite triennial — the survey which should have covered 1987 on a strict triennial basis now refers to 1988. Information on property transactions is very unreliable. It comes from a variety of sources including exchange control information dating from 1979. Fuller details are given in chapter 5.

4.2.2 *Portfolio Investment*

Overseas investment by UK residents is given in the balance of payments [QRL.61] for four categories of investors: UK banks, other UK financial institutions, other UK residents, all UK residents. The investment of these categories of investor is given by type of instrument: (1) bonds, preference shares, debentures etc. (2) ordinary shares, and total.

Overseas investment in UK securities by overseas residents is given by type of UK instrument. These are (1) British government stocks (investment by overseas authorities, exchange reserves, other overseas residents, total), (2) British government foreign currency bonds and notes, (3) local authorities' securities (sterling, foreign currency under the exchange cover scheme, other foreign currency), (4) public corporations' securities (issued under the exchange cover scheme, other (foreign currency), (5) UK companies' securities (issued under the exchange cover scheme, other bonds etc, ordinary shares). The table (8.3 in the 'Pink Book') also shows total investment in securities issued by general government, public corporations, UK companies.

Portfolio investment, covering all securities with an original contractual maturity of more than one year, as well as corporate equities, bonds and other securities with no specific maturity excludes investment in short-term instruments (eg Treasury bills and commercial paper) which are included in 'other capital transactions'. Because of the diverse holding of such assets, basic information is derived from a variety of sources, including banking statistics and Bank of England sources for foreign investment in the UK. Information on UK residents' investment abroad is derived from a Bank of England statistical enquiry to UK dealers in securities and banks together with enquiries to UK non-bank financial institutions. Estimates are made for transactions not reported directly. It is likely that the statistical net is here (as elsewhere) not wide enough to capture all the transactions. The direction of the error is not known, although it may be guessed that underestimation is the order of the day.

4.2.3 *Transactions of UK Banks*

Transactions of UK banks are shown both for transactions in external assets of the UK and in UK liabilities to overseas residents.

Transactions in external assets of the UK are (1) foreign currency lending abroad by UK banks (identified long-term export credit, other foreign currency lending) and (2) sterling lending abroad by UK banks (identified long-term export credit, other identified export credit, other sterling lending).

Transactions in UK liabilities to overseas residents are foreign currency borrowing and deposit liabilities abroad of the UK banks plus sterling borrowing and deposit liabilities abroad of UK banks (overseas authorities' exchange reserves in sterling, other liabilities).

Net transactions are shown in foreign currency and in sterling.

Information on banks abroad is unreliable. For instance, information on transactions with banks abroad, specifically their level of bank deposits is compiled from World Bank sources but it contains discontinuities and breaks.

4.2.4 *Transactions of UK Residents Other than Banks and General Government.*

This residual category of transactions in assets and liabilities examines transactions in external assets of the UK and UK liabilities to overseas residents of the UK non-bank private sector and public corporations.

Transactions in UK external assets cover (1) deposits and lending overseas (transactions with banks abroad by the UK non-bank private sector) plus (2) other assets of the UK non-bank private sector and public corporations. The latter assets are: long-term capital transactions of the Commonwealth Development Corporation, advance and progress payments on imports (long-term agreements, other), export credit, other lending and short-term assets, foreign notes and coin. The total is divided between public corporations and the UK non-bank private sector.

Transactions in UK liabilities to overseas residents covers borrowing from overseas plus other liabilities. Borrowing from banks abroad is given for public corporations and for the UK non-bank private sector (identified long-term import credit and other). Other liabilities are: import credit (long-term liabilities and other), identified advance and progress payments on exports and other (mainly short-term) liabilities. Total transactions in liabilities of the two groups are also given under the exchange cover scheme and other for the UK non-bank private sector. Net transactions are similarly apportioned.

4.2.5 *Transactions of General Government*

The transactions of general government are shown excluding portfolio investment liabilities (see section 4.2.3) [QRL.61].

Transactions in external assets of the UK cover official reserves plus other external assets of central government. The other external assets are: inter-government loans by the UK, other long-term capital transactions viz subscriptions to international lending bodies (International Development Association, Regional Development Banks, European Investment Bank, other subscriptions) and other long-term assets, export credit and official short-term capital.

Transactions in UK liabilities to overseas residents cover external liabilities of general government other than portfolio liabilities: (1) transactions with the IMF (UK drawings, re-purchases, other net drawings and IMF expenditure and receipts in sterling), (2) net transactions with other monetary authorities: sterling and non-sterling borrowing by the UK, (3) inter-governmental loans to the UK, (4) foreign currency borrowing from banks etc. overseas (by HM Government, Northern Ireland central government, local authorities under exchange cover scheme and other local authority), (5) sterling borrowing from, and deposit liabilities to, banks etc. overseas by local authorities, (6) Treasury bills and non-interest bearing notes issued by HM Government, (7) official short-term capital transactions and sterling notes and cover held by private (non-bank) overseas residents. The total transactions in liabilities of central government are given as well as the total for local authorities and Northern Ireland central government, as are net transactions in assets and liabilities of general government. The allocation of SDRs to the UK by the IMF is shown separately from these totals.

4.2.6 *Reliability*

The comments on the general reliability of transactions in UK external assets and liabilities are to be found in section 1.3.5 above.

4.3 Levels of UK External Assets and Liabilities

Strictly speaking, levels of external assets and liabilities are stocks and not flows and so have little direct influence on the balance of payments. This section of the balance of payments accounts [QRL.61] shows the effect of transactions on the levels of UK external assets and liabilities at year-end. Changes in the value of these stocks show not only the effect of annual capital transactions but also revaluations and other changes. These levels can be compared with transactions figures (section 4.2 above) and earnings (section 3.3 above) although the coverage, in particular of short-term assets and liabilities, is incomplete. The 'Pink Book' [QRL.61] contains a warning that there are many inconsistencies in the tables showing levels of external assets and liabilities because of the variety of data sources used and, in particular, problems in the methods of valuation and the unidentifiability of certain assets and liabilities. This should be at the forefront of any user's mind.

4.3.1 *Valuation and Coverage*

Wherever possible market values are used. Revaluations occur with (1) changes in *book* or market values (direct investment is recorded at book values and book values do not always match recorded flows because of exchange rate changes, revaluations and disposals at more or less than book values), (2) changes in sterling values of foreign-currency-denominated assets or liabilities consequent on exchange rate changes, (3) in principle, creation or extinction of assets and liabilities such as writing-off bad debts, where there is no associated flow of investment (clearly this is difficult to identify and in practice can rarely be accurately included), (4) coverage changes resulting from modified reporting systems (e.g. UK banks end-1974, UK monetary sectory end-1981), (5) coverage differences resulting from changes in the population of reporting institutions.

These factors make valuation extremely difficult. This is particularly true where new financial instruments are involved and non-standard assets and liabilities are being estimated.

4.3.2 *Direct Investment*

A sector analysis only of levels of direct investment is included in the 'Pink Book' [QRL.61] — compare 4.2.1 above. Raw data comes from a periodic DTI survey (see 5.2.11 below) with annual figures interpolated. Direct investment stocks are covered in detail in Chapter 5.

4.3.3 *Portfolio Investment*

UK assets since 1980 onwards have been calculated using a combination of banking statistics, the results of the Bank of England's portfolio enquiry to dealers in securities and information (available or estimated) from financial institutions and

Lloyd's together with revaluation factors applied to holdings by other investors. The comments on accuracy in 4.2.2 apply equally here.

4.3.4 *Assets and Liabilities of UK Banks.*
This is covered in section 7.9 below as regards balance of payments data, and much more fully in Chapter 7 *passim.*

4.3.5 *Assets and Liabilities of General Government*
The presentation of levels of assets and liabilities of general government in the *UK Balance of Payments* [QRL.61] follows that of annual transactions as in 4.2.5 above. Revaluation of such levels may be frequent as inter-government loans are 'forgiven' (i.e. written off), although the practice in recent years has been to regard payments of capital and interest as being made on schedule. Amounts equal to these repayments are then included as transfer debits: bilateral non-project grants in the transfers section of invisible trade (see section 3.5 above).

General government assets and liabilities are usually measured from official records and can be regarded as reliable.

4.4 Financial Accounts and Balance of Payments Accounts

The *United Kingdom National Accounts*, the 'Blue Book' [QRL.64] also gives financial accounts for the overseas sector. This is reconciled with the *United Kingdom Balance of Payments* [QRL.61] accounts according to the Notes and Definitions supplement to *Financial Statistics* [QRL.33].

4.5 Financial Statistics Data

Financial Statistics [QRL.33] is a useful summary source of transactions of the overseas sector and for exchange rates and interest rates. The annual *Explanatory Handbook* gives full descriptions of the sources of the basic data, together with some indication of its quality and deficiencies. Users are advised to check the notes to individual tables in the *Explanatory Handbook*

Information on the overseas sector (currently section 10 of *Financial Statistics*) cover the current account (on a less detailed basis than *The UK Balance of Payments* [QRL.61] which is the primary source) and transactions in UK external assets and liabilities. Official reserves, external liabilities and claims in sterling, external liabilities and claims of the United Kingdom in foreign currencies and outstanding government and other foreign currency borrowing are covered in four separate tables. One table covers the International Monetary Fund (IMF) and presents the current UK position with the IMF and UK transactions with the Fund. Finally, transactions in overseas securities by central government, monetary sector, other financial institutions, industrial and commercial companies and the personal sector are shown together with the overseas sector capital issues and other transactions. *Financial Statistics* [QRL.33] also includes information on foreign exchange rates which are fully examined in chapter 8.

A reconciliation between the financial accounts and the balance of payments is published as a supplementary table which last appeared, with figures for 1986, in the October 1987 issue of *Financial Statistics* [QRL.33]. The system of sector accounts is shown in a table with explanatory notes, in the *Explanatory Handbook*.

CHAPTER 5

OUTWARD AND INWARD DIRECT INVESTMENT—ANNUAL TRANSACTIONS AND BOOK VALUES

5.1 Introduction and Definitions

5.1.1 *Introduction*

It is the purpose of this chapter to review the available data on UK direct investment overseas and foreign direct investment in the UK. With the exceptions of sections 5.6 to 5.8, this involves a focus on the detailed information deriving from a series of surveys carried out over the past quarter century by the Department of Trade and Industry and its predecessors. There have been two series of these surveys. An annual series (discussed in sections 5.3 and 5.5) has collected information on 'flows' of new direct investment and on earnings on direct investment. The other series (discussed in sections 5.2 and 5.4), mainly triennial, has compiled data on the stocks of direct investment existing at points in time. The form taken by stock figures is that of 'book values' (see 5.2.1.2), this is generally regarded as conceptually less satisfactory than market value or replacement cost data, but the only form where efficient data collection is viable.

5.1.2 *Direct Investment—General Sources*

5.1.2.1

It should be noted that though the sources surveyed in depth here provide extensive detail on direct investment (by host or investing country; by industry; by component) other sources provide information on direct investment at a more aggregate level, usually placed in some broader context.

5.1.2.2

The 1986 edition of the *UK Balance of Payments* 'Pink Book' [QRL.61] for example, provides data on inward and outward direct investment in the capital account in table 1.4 (see also table 8.1), with this subdivided by components in table 8.2 and by sector in table 8.2A. Annual direct investment earnings, in the context of all earnings on UK external assets and liabilities, are also provided in table 5.1 in the 1986 edition of *UK Balance of Payments* [QRL.61], with subdivision by components (table 5.2) and by sectors (table 5.2A). Annual estimates (see 5.2.1.1) of book values of outward and inward direct investment are included, in the context of total external assets and liabilities, in table 9.1 in the 1986 edition of *UK Balance of Payments* [QRL.61], with subdivision by sector in table 9.2A.

5.1.2.3
Information on outward and inward direct investment assets is also included in an annual article in the *Bank of England Quarterly Bulletin* [QRL.4], which surveys developments in the UK's external balance sheet. Thus, for example, table G in the article contained in the September 1986 issue places bank and non-bank direct investment assets and liabilities in the wider external balance sheet context.

5.1.2.4
Data produced in quarterly (June, September, December, March) balance of payments articles in *Economic Trends* [QRL.30] also provide aggregate figures on direct investment, including quarterly estimates for recent years. Thus direct investment earnings, in the context of total 'interest, profits and dividends', are covered in table 6 of the March 1987 issue. The quarterly earnings figures in this table (covering 1982 to 1986) are presented in a seasonally adjusted form, as well as unadjusted. Table 8 in the March 1987 *Economic Trends* [QRL.30] puts changes in inward and outward direct investment assets in the context of total transactions in external assets and liabilities. This issue provided data annually for 1975 to 1986 and quarterly (but not seasonally adjusted) from 1978 to 1986. Further detail on direct investment transactions are provided in table 9 of March 1987 *Economic Trends*.

5.1.3 *Definitions of Direct Investment*

5.1.3.1
Direct investment is defined in the *IMF Balance of Payments Manual* [B.26] as "investment that is made to acquire a lasting interest in an enterprise operating in an economy other than that of the investor, the investor's purpose being to have an effective voice in the management of the enterprise". A subsequent OECD report seeks to provide a more precise operational definition of direct investment, in line with the principle set down by the IMF. This OECD report, and UK direct investment reporting practices in the light of the OECD standards, are described by Walker [B.24].

5.1.3.2
The OECD recommendation is that "a direct investment enterprise be defined as an incorporated or unincorporated enterprise in one country that either has a foreign investor controlling 10 per cent or more of its equity capital or voting power, unless this is insufficient to allow the investor to participate in the enterprise's management, or has a foreign investor with an effective voice in the enterprise's management even if the investor controls less than 10 per cent of the equity or voting power" (Walker; [B.24]; p. 61.1). So direct investment, by contrast to portfolio investment where the motivation is limited to financial gain, seeks to achieve operating control over the enterprise in which the investment is made. Further detail of the coverage of direct investment can be found for book value data in 5.2.1.2 and for annual direct investment in 5.3.1.3.

5.1.4 *UK Applications*

It is an increasingly important facet of the spread of multinational companies that overseas subsidiaries often themselves have further subsidiaries in third countries. This provides certain conceptual problems in the compilation of direct investment data (see Walker [B.24]; pp. 61.3-61.4).

The treatment of inward direct investment in the UK seems to be clearcut, with investment and earnings attributed to the country of the immediate investing affiliate, rather than the country of the ultimate owning parent. Table 25 in the report of the 1981 book value survey ([QRL.10]; 1981 Supplement) provides inward asset data by country of immediate parent and country of ultimate parent.

Walker ([B.24] p. 61.4), however, describes the treatment of outward statistics as a mixture of ultimate and immediate countries. It is implied that with the increasing availability of statistics for both directly and indirectly held subsidiaries, the UK outward country figures "are fairly close conceptually" to those of the USA, which allocates its outward flows, earnings and stock of investment to the country of the indirectly owned unit. Table 13 in the report of the 1981 book value survey ([QRL.10]; 1981 Supplement) provides outward asset data by 'direct investment country' and 'immediate investment country'.

5.1.5 *Organisation of Area Coverage*

The way in which the area coverage of UK direct investment data is organised has varied over time for a number of reasons.

Firstly, the formal changes in the composition of the EEC and EFTA have been reflected in the area coverage.

Secondly, the coverage up to 1974 for annual transactions data, and up to 1971 for book value data, was organised around a distinction between 'overseas sterling area' countries and 'non-sterling area' countries. The use of this distinction was discontinued after those dates.

Thirdly, the *Board of Trade Journal* [QRL.5] article dated 19th July 1968 adopted the OECD definition of 'developing countries', rather than the UN definition which had been previously used. At the time this resulted in Spain, Greece, Turkey and Yugoslavia becoming classified as developing countries.

5.2 UK Direct Investment Overseas—Book Values

5.2.1 *Introduction and Definitions*

5.2.1.1

The Department of Trade and Industry, and its predecessors, carried out inquiries into the book values of direct investment (both UK investment overseas and foreign investment in UK) for end-years 1984 (see Appendix 7), 1981, 1978, 1974, 1971, 1968, 1965 and 1962. These inquiries aimed to cover all companies except those in oil, banking and insurance. As we shall see incorporation of these sectors too has gradually become possible.

Since 1978 the DTI inquiry into UK overseas direct investment has been on a statutory basis, though the comparable inquiry into inward direct investment (section 5.4 below) remains voluntary.

The results of these inquiries have been reported in a series of *Business Monitors M4/MA4* [QRL.10], and before the introduction of the *Business Monitor* series in the *Board of Trade Journal* [QRL.5]. Since the inauguration of the *Business Monitor* series less detailed coverage of the asset surveys have continued to be published in the *Board of Trade Journal* [QRL.5], and its successors, *Trade and Industry* [QRL.60] and *British Business* [QRL.7]. Details of these sources are listed in Appendix 5.

It should be noted that the *UK Balance of Payments*, the 'Pink Book', [QRL.61] includes in its external assets and liabilities section (Tables 9.1 and 9.2A in the 1986 edition) annual estimates of book values of direct investment assets, i.e. including interpolated values for years other than those listed above as covered by detailed inquiries. The *Business Monitor MA4* [QRL.10], however, draws attention to the inevitably tentative nature of these interpolated estimates. The change in the book value each year consists of the new investment made in that year, plus changes in the sterling value of foreign currency assets and liabilities which may result from exchange rate movements or from revaluations of assets, whilst in recent years changes in the method of accounting for deferred tax has had the effect of increasing the value of net assets due to the direct investor. The data collected annually on new investments in that year (described in 5.3 below) help considerably with the calculation of these interpolated book values, but estimates have to be made for the other factors based on foreign currency movements and past experience.

5.2.1.2

The net book value of direct investment is the value of a company's investment in its overseas subsidiaries, branches and associates.

The net book value of direct investment in *subsidiaries* is defined as the fixed assets of the subsidiary (net of depreciation), plus the subsidiary's current assets less current liabilities excluding those due to or from the investing company, less subsidiary's long-term loans due to third parties, less deferred tax reserves and provisions, and less minority shareholders' interests in the subsidiary. This then equals the net assets due to the investing company, which in turn equals the proportion of the subsidiary's share capital and reserves due to the investing company, plus the subsidiary's long-term loans and current liabilities due to the investing company, less the subsidiary's current assets and loans due from the investing company.

The net book value of direct investment in *branches* are defined as the fixed assets of the branch (net of depreciation), plus the branch's current assets less its current liabilities (excluding those due from or to the investing company), less the branch's long-term loans due to third parties, and less deferred tax reserves and provisions. This equals the net assets of the branch due to the investing company.

UK investment in overseas *associates* is defined as the book value of the investment as shown in the accounts of the UK investing company, plus loans and net current account balances due to the UK investing company from the overseas associate. The book value in the investing company's accounts is usually the cost of the share capital of the associate when purchased less any amounts written off, plus the investing company's share of the post acquisition retained profits and reserves of the associated company.

The *Business Monitor MA4* [QRL.10] draws attention to the fact that much of the UK direct investment overseas is financed by foreign currency borrowing by UK companies from UK and overseas banks, and notes that the direct investment statistics include the direct investment financed by this borrowing, but not the corresponding liability to the lender of the foreign currency. On the other hand where a branch, subsidiary or associate borrows money itself from local banks or other third parties, this borrowing is omitted since it does not increase the value of the net assets attributable to the investing company.

5.2.2 *Country Data*

5.2.2.1

The census of overseas assets for end-1981 ([QRL.10]; 1981 Supplement) is the first to produce (Table 1) book value data for all industries (i.e. including oil, banking and insurance). This achievement of complete coverage was facilitated by the availability in 1981 of data from a newly-implemented survey of the book value of the direct investment overseas of UK insurance companies. This survey was carried out by the British Insurance Association. The table provides quite extensive coverage at the country level, but further information could not be given without disclosing detail of individual oil, banking and insurance companies.

5.2.2.2

The census of overseas assets for end-1978 ([QRL.10]; 1978 Supplement) had been the first to produce book value data for all industries other than insurance (i.e. incorporating oil and banking for the first time). In fact, the Bank of England had collected overseas asset data for oil companies for some time, but could not publish figures for individual countries without disclosing information about individual companies. However, in 1977 the Bank of England started to collect information on the book value of overseas direct investment of banks. As a result of this, in 1978, it was possible to combine the information on banks and oil companies in a way that avoided disclosure, thus enabling figures for all industries except insurance to be published, including country level coverage. Some of this data was subsequently revised and the definitive revised data for 1978 is produced as table 1A in the *Business Monitor MA4* for 1981 ([QRL.10]; 1981 Supplement).

Data on overseas assets of UK oil companies (and foreign oil companies in the UK) have been published, without country breakdown, in the *UK Balance of Payments* 'Pink Book' ([QRL.61]; 1985 edition; tables 11.1 and 11.2). The 1986 edition of the 'Pink Book', however, discontinued separate presentation of direct investment data for oil companies.

5.2.2.3

The basic source for a time series of data on book values of net assets of UK overseas direct investment (excluding oil, banking and insurance), in considerable country and area detail, is table 2 in the 1981 *Business Monitor* MA4 ([QRL.10]; 1981 Supplement). This includes data for each of the years covered by the asset inquiries (see 5.2.1.1). Similar data was, of course, presented in each of the earlier *Business Monitor* and *Board of Trade Journal* sources reporting asset data (see Appendix 5), but the most recent table essentially supersedes them. An exception to this which should be noted is that some of the earlier sources report interpolated figures for non-inquiry years. Notably the *Board of Trade Journal* article reporting the 1965 book value survey ([QRL.5]; 26th January 1968) provides interpolated estimates for 1960 and 1961, i.e. working back from the first inquiry in 1962. The difficulties of this procedure were noted in 5.2.1.1.

5.2.3 *Industry Data (excluding Oil, Banking, Insurance)*

5.2.3.1

A time series of data dividing overseas direct investment assets into 'manufacturing' and 'non-manufacturing' appears in the 1981 *Business Monitor MA4* ([QRL.10]; 1981 Supplement; Table 4). This covers all the years of book value inquiries, except 1962. The area coverage is quite broad, comprising:

(1) Western Europe; subdivided into EC and EFTA
(2) North America; subdivided into Canada and USA
(3) Other Developed Countries
(4) Rest of World.

5.2.3.2

Each of the overseas asset inquiries, except that for 1962, has produced a matrix table of data for book values by country and by industry of overseas affiliate. Thus the table for 1981 ([QRL.10]; 1981 Supplement; table 6) covers 11 manufacturing and 8 non-manufacturing industries (see Appendix 2). The industry coverage for 1978 was the same as for 1981, whilst that for 1974 differs only by providing separate data for 'shipping' and 'transport and communications', which were combined in the subsequent years. The industry coverage for 1971 is as for 1974, except that 'shipbuilding' was not covered separately. The industry coverage for 1965 and 1968 is less detailed, covering 10 industries in total with manufacturing divided into three sectors only. In addition to a certain amount of change in the detailed country coverage, the area organisation of these tables alters in line with the principles noted above (section 5.1.5). It should be noted that the 1968 data here is available from the *Board of Trade Journal* source ([QRL.5]; 23rd September 1970) but not the equivalent *Business Monitor*.

5.2.3.3

The possibility that the industry of operation of an overseas affiliate may differ from that of its UK parent is documented in table 7 of the 1981 inquiry ([QRL.10]; 1981 Supplement). This table provides a matrix (covering 19 industries) for the book value of direct investment assets by industry of UK enterprise and industry of overseas affiliate. A similar table is available for 1978, and (with somewhat different industry coverage) for 1974 and 1971. It should be noted that, unlike the subsequent years, the table for 1971 is limited to an analysis of the returns received in the survey, with no attempt to provide estimates for missing coverage.

5.2.4 *Forms of Organisation*

5.2.4.1

Data from the 1981 inquiry ([QRL.10]; 1981 Supplement; table 5) provide information, with detailed area and country breakdown, on the organisational form taken by UK overseas direct investment. This investment is divided into:

(1) Subsidiaries: (a) number, (b) book value
(2) Branches: (a) number, (b) book value
(3) Associates: (a) number, (b) book value
(4) Total: (a) number, (b) book value

Data on the same basis are also provided from the 1978 and 1974 inquiries.

5.2.4.2

Similar data to that for 1974 and 1981 were derived in the 1971; 1968; 1965 inquiries. Here investment is divided into:

(1) Large investing companies—Subsidiaries (a) number (b) book values
(2) Large investing companies—Branches (a) number (b) book values
(3) Large investing companies—Associates (a) number (b) book values
(4) Small investing companies (a) number (b) book values

The limit for small investing companies was £1 million in 1971, £0.5 million in 1968, and £0.25 million in 1965. The data for 1968 is available from the *Board of Trade Journal* source ([QRL.5]; 23rd September 1970), but not from the equivalent *Business Monitor*.

5.2.5 *Percentage Shareholding*

5.2.5.1

Data from the 1981 inquiry ([QRL.10]; 1981 Supplement; table 8) provides information, with a detailed area and country breakdown, on UK percentage shareholding in overseas direct investment subsidiaries. The ownership categories (i.e. UK percentage shareholding in subsidiary) covered are:

(1) Less than 50% (Number of subsidiaries; book value)
(2) 50% to 74.9% (Number of subsidiaries; book value)

(3) 75% to 99.9% (Number of subsidiaries; book value)
(4) 100% (Number of subsidiaries; book value)
(5) Total (Number of subsidiaries; book value)

The book values referred to are the overseas assets *owned* by UK investors in subsidiaries falling in a particular ownership category. For the less than 100% ownership categories this will thus be less than the full value of these subsidiaries, which, in line with the definition of direct investment would be considered to be *controlled* by the UK investor. Similar data had been reported from the 1978 and 1974 inquiries.

5.2.5.2

The 1981 inquiry ([QRL.10]; 1981 Supplement; table 9) also performed an analysis of ownership patterns of overseas associates comparable to that for subsidiaries described in the previous section. The ownership categories (i.e. UK percentage shareholding in associates) covered are:
(1) Less than 20% (number of associates; book value)
(2) 20% to 34.9% (number of associates; book value)
(3) 35% to 49.9% (number of associates; book value)
(4) 50% or more (number of associates; book value)
(5) Total (number of associates; book value)

Detailed area and country coverage is provided. Similar data have also been reported from the 1974 and 1978 inquiries.

5.2.5.3

Since branches are, by definition, 100 per cent owned these are not covered in the same way as subsidiaries and associates described above.

5.2.5.4

Some information on ownership patterns was derived from the 1965 inquiry. Thus table 8 ([QRL.5]; 26th January 1968) provides information in the categories listed in 5.2.5.1 above, but for subsidiaries and associates combined. Both numbers of affiliates and book values are covered, but the area coverage is restricted to overseas sterling area/non-sterling area and developed countries/developing countries. Table 9 provides similar information on ownership patterns with an industry breakdown.

5.2.6 *Size of UK Enterprises' Direct Investment*

5.2.6.1

Data produced by recent inquiries provides some indication of the nature of distribution of UK overseas direct investment assets among UK enterprises. Thus in the report of the 1981 survey ([QRL.10]; 1981 Supplement) table 12 first of all divides UK parent companies into 10 size groups, according to the value of their total overseas assets. These size groups range from total overseas net assets 'under £1 million' to 'over £200 million'.

For each size group data are then given on:

(1) Number of UK enterprises
(2) Number of overseas subsidiaries
(3) Number of overseas branches
(4) Number of overseas associates
(5) Number of overseas affiliates (i.e. (2) + (3) + (4))
(6) Book value of net assets of overseas affiliates.

5.2.6.2

Data similar to that described in 5.2.6.1 is available for 1978, and for 8 size groups ('under £1 million' to 'over £100 million') for 1974. Less detailed coverage is also available for 1971, where 7 size groups are covered, with data for each on:

(1) Number of UK enterprises
(2) Number of overseas branches, subsidiaries and associates
(3) Book value of net assets of overseas affiliates.

5.2.7 *Balance Sheet Analysis of Overseas Subsidiaries*

5.2.7.1

An important part of the detailed information on overseas investment provided by the series of book value inquiries has been balance sheet analysis of subsidiaries assets and liabilities. Here we review the 1981 coverage in detail, and draw attention to some of the changes in presentation in earlier inquiries. In all years the data are given with a detailed area/country coverage.

5.2.7.2

The balance sheet analysis for subsidiaries for end-1981 ([QRL.10]; 1981 Supplement; Table 10) provides data on:

(1) *Net assets*; divided as
    (i) Fixed assets (after provision for accumulated depreciation), plus
    (ii) Total current assets, less
    (iii) Total current liabilities

In addition 'total current assets' are divided as (a) due from UK parent companies, plus (b) other. Similarly 'total current liabilities' are divided as (c) due to UK parent companies, plus (d) other.

(2) *Total share capital and reserves*; divided as
    (i) Net assets, less
    (ii) Total long term liabilities, less
    (iii) Minority shareholders interests in subsidiaries consolidated in overseas companies accounts.

In addition 'total long-term liabilities' are divided as (a) those due to UK parent companies, plus (b) other long-term liabilities, plus (c) provision for future taxation.

(3) *Total share capital and reserves*; divided as
 (i) attributable to minority interests, plus
 (ii) attributable to UK parent companies
(4) *Total book value of net assets attributable to the UK*; divided as
 (i) share capital and reserves attributable to UK parent companies, plus
 (ii) net current and long-term liabilities equivalent to net assets in the books of UK parent companies.

5.2.7.3

The 1978 subsidiaries balance sheet coverage ([QRL.10]; 1978 Supplement; Table 10) is essentially that described for 1981 above. Two variations may be noted, however. Firstly, 'fixed assets' are subdivided as (a) fixed assets before provision for accumulated depreciation, less (b) accumulated depreciation. Secondly, data are not given on 'minority shareholders interests' (i.e. (2) (iii) above).

5.2.7.4

The 1974 coverage ([QRL.10]; 1974 Supplement; Table 8) is as for 1978 except that 'provision for future taxation' is included in 'current liabilities' rather than in 'long term liabilities'.

5.2.7.5

The organisation of subsidiaries balance sheet data for 1971 ([QRL.10]; 1972 edition part 11; Table 38) differs from that in subsequent years described above. Data are provided on:
(1) *Total net assets*; divided as
 (i) total assets, less
 (ii) total liabilities
(2) *Total net assets*; divided alternatively as
 (i) those attributable to outside shareholders, plus
 (ii) those attributable to UK parent companies
(3) *Total assets*; divided as
 (i) fixed assets after provision for depreciation, plus
 (ii) total current assets, less
 (iii) that part of total current assets due from UK parent
(4) *Total liabilities*; divided as
 (i) long term liabilities overseas, plus
 (ii) provision for future taxation, plus
 (iii) total current liabilities, less
 (iv) that part of total current liabilities due to UK parent
(5) *Accumulated provision for depreciation of fixed assets*

It should be noted that the data for 1971 are based on the analysis of 'returns received' in the survey, without (unlike subsequent years) any estimated allowance for non-response.

5.2.7.6
The coverage for 1968 ([QRL.10]; 1969 edition; Table 35) is as for 1971, and also, like that year, is based on 'returns received' only.

5.2.7.7
The data for 1965 ([QRL.5]; 26th January 1968; Table 5) are limited to 'large investing companies', but (unlike 1968 and 1971) does provide allowance for non-response. Otherwise the coverage is as for 1968/1971, except that 'long-term liabilities overseas and provisions for future taxation' are not separated as later (i.e. (4) (i) and (4) (ii) in 5.2.7.5 above).

5.2.7.8
The first asset inquiry ([QRL.5]; 7th August 1964; Table 1) provided balance sheet data for overseas subsidiaries and branches taken together. The format is a less detailed equivalent to that described in 5.2.7.5. The detailed figures are based on 'returns received', without allowance for non-response.

5.2.8 *Balance Sheet Analysis of Overseas Branches*

5.2.8.1
The various asset inquiries have collected balance sheet data for overseas branches in a manner similar to that for subsidiaries. Here we provide a survey of these data comparable to that for subsidiaries in 5.2.7 above. In each year the data are given with a detailed area/country coverage.

5.2.8.2
The balance sheet analysis for branches for end-1981 ([QRL.10]; 1981 Supplement; Table 11) provides data on:
(1) *Net assets*; divided as
 (i) fixed assets (after provision for accumulated depreciation), plus
 (ii) current assets, less
 (iii) current liabilities
(2) *Book value of net assets attributable to UK companies*, divided as
 (i) net assets, less
 (ii) long term liabilities, less
 (iii) provision for future taxation

5.2.8.3
The 1978 coverage for branches ([QRL.10]; 1978 Supplement; Table 11) is as for 1981, except that 'fixed assets' is subdivided as (a) fixed assets before provision for accumulated depreciation, less (b) accumulated depreciation.

5.2.8.4

The 1974 coverage ([QRL.10]; 1974 Supplement; Table 9) is as that for 1978 except that 'provision for future taxation' is included in 'current liabilities' rather than as a long-term component i.e. (2) (iii), in 1978.

5.2.8.5

The organisation of branch balance sheet data for 1971 ([QRL.10]; 1972 edition part 11; Table 39) differs from that in subsequent years described above. Data are provided on:

(1) *Total net assets*; divided as
 (i) total assets, less
 (ii) total liabilities

(2) *Total assets*; divided as
 (i) fixed assets after provision for depreciation, plus
 (ii) current assets

(3) *Total liabilities*; divided as
 (i) long-term liabilities overseas, plus
 (ii) provision for future taxation, plus
 (iii) current liabilities

(4) *Accumulated provision for depreciation of fixed assets*

Data for 1971 are based on the analysis of 'returns received', without the allowance for non-response made in subsequent years.

5.2.8.6

The coverage for 1968 ([QRL.10]; 1969 edition; Table 36) is as for 1971 and like that year is based on 'returns received' only.

5.2.8.7

The data for 1965 ([QRL.5]; 26th January 1968; Table 6) are limited to 'large investing companies', but (unlike 1968 and 1971) do provide allowance for non-response. Otherwise the coverage is as for 1968 and 1971, except that 'long-term liabilities overseas and provision for future taxation' are not separated as later (i.e. (3) (i) and (3) (iii) in 5.2.8.5 above).

5.2.9 *Date of Establishment Data (1965 Inquiry)*

5.2.9.1

The asset inquiry for 1965 ([QRL.5]; 26th January 1968), derives some information providing useful hints as to how UK direct investment overseas had evolved in the period before the systematic collection of detailed data. This consists of dividing the large investing companies surveyed at end-year 1965 by the period in which they were originally established. The time periods covered for original establishment are

(1) before 1946, (2) 1946-55, (3) 1956-65. Thus the asset values reported in the 'date of establishment' tables are those for net assets attributable to UK at end-year 1965, but categorised according to the date of original establishment of the overseas operation. They do not represent an age distribution of the assets employed at end-1965.

5.2.9.2
The most detailed table reported in the *Board of Trade Journal* article ([QRL.5]; 26th January 1968) on this basis (Table 10), provides data for period of establishment by industry (10 industries, of which 3 are in manufacturing) and by form of organisation (subsidiaries; branches; associates). In addition to the 1965 assets in these categories this table also reports data on numbers of firms. It is important to underline, as described for assets above, that the figures for numbers of firms relate to those existing at end-1965 not all those originally established in each of the three time periods.

The other detailed table incorporating period of establishment figures (Table 7) provides this information by form of organisation and with an extensive area and country coverage.

5.2.9.3
Two smaller tables, incorporated in the text of the *Board of Trade Journal* article, relate date of establishment to (a) size of investment at end-1965, and (b) degree of UK ownership of the investment at end-1965.

5.3 UK Direct Investment Overseas—Annual Transactions

5.3.1 *Introduction and Definitions*

5.3.1.1
The transactions covered by the annual surveys may be divided into those relating to new direct investment overseas (see sections 5.3.2 to 5.3.4), and those relating to the earnings of UK overseas direct investment (see sections 5.3.5 to 5.3.8). These sets of transactions are not, of course, mutually exclusive; most notably both include unremitted profits.

5.3.1.2
The coverage of the UK outward direct investment statistics (see Walker [B.24]; pp. 61.1-61.2) includes all overseas incorporated enterprises that are subsidiary or associate companies of UK concerns, plus all overseas unincorporated businesses or branches owned directly by UK concerns. Here a subsidiary is a company in which the UK concern either directly owns more than half the equity share capital or controls the board of directors, or the company is a subsidiary of another company

which is itself a subsidiary of the UK concern. Generally an associate is a company in which between 20 and 50 per cent of the share capital is held by the UK concern and/or its overseas subsidiaries. Walker [B.24] notes that there are a few exceptions to this where the UK concern controls less than 20 per cent but is involved in the overseas companies management, and by contrast a few exclusions where more than 20 per cent is held as a portfolio investment. The UK definition then differs from the OECD guideline (see section 5.1.3.2) by using a cut off of 20 per cent of equity instead of 10 per cent, though the amount of investment falling in this range is likely to be small.

5.3.1.3

Direct investment (see sections 5.3.2 to 5.3.4) consists of the parent company's share of the unremitted profits, less losses, of subsidiary and associated companies; the cash acquisition, net of sales, of share and loan capital of subsidiaries and associated companies; changes in indebtedness on inter-company accounts and changes in branch–head office indebtedness.

The inter-company account between the parent and its subsidiaries and associates, and the branch–head office account, cover amounts due to the parent from, less amounts due from the parent to, its related overseas concerns for short-term loans, trade credit and any other indebtedness not covered elsewhere.

This coverage is essentially in line with the OECD guidelines (see Walker [B.24]; pp. 61.2-61.3) on definition of direct investment (see sections 5.1.3.1).

5.3.1.4

The earnings on direct investment (see sections 5.3.5 to 5.3.8) consists of the profits of branches; the parent company's share of the profits of subsidiaries and associates; and interest payments by subsidiaries and associates to, less payments from, the parent company on inter-company loans.

Profits include unremitted profits as well as those remitted to the parent, and are net of losses and measured after deducting provisions made for depreciation of fixed assets and provisions for overseas tax. The UK's treatment of direct investment earnings thus adheres closely to the OECD guidelines, in an area of coverage where quite a few leading countries do not (see Walker [B.24]; pp. 61.2).

5.3.1.5

It should be noted that the annual overseas direct investment figures surveyed below (5.3.2 to 5.3.4) are normally *net*, in that they represent the difference between gross direct investment overseas less gross disinvestment from overseas. Up to the 1981 *Business Monitor MA4* ([QRL.10] 1981 edition; table 1.6) data had been given on gross investment and gross disinvestment, including analysis by component.

5.3.1.6
In the coverage of annual outward direct investment transactions most industries have been incorporated since the initiation of the surveys. Insurance, however, was omitted from 1960 to 1962. Coverage of oil companies' direct investment was included in 1984 for the first time. It should be noted that in the detailed industry breakdown incorporated in some tables the inclusion of petroleum in 1984 is achieved not as a separate new industry, but as part of the existing 'mining' industry, which becomes 'mining and petroleum', by adding SIC classes 13 and 14 (see Appendix 2). The absence of independent 1984 figures for 'petroleum' means that it is not possible to adjust these data to achieve comparability with earlier years at the aggregate level. A consistent time series for 'manufacturing' only, however, remains feasible.

5.3.1.7
Though information derived from the outward direct investment surveys has been published since 1960 (see Appendix 6), we can delineate a selected number of 'main sources' for this data. These are the sequence of *Business Monitors M4/MA4* [QRL.10] dated 1969 to 1984, plus the *Board of Trade Journal* [QRL.5] articles published in 30th June 1967, 19th July 1968, 9th May 1969 editions. Most of the information in earlier sources is included in time series in these 'main sources'.

5.3.1.8
It should be noted that when table numbers are quoted for construction of a time series (as, for instance, in section 5.3.2.1), the years referred to are the current year of *data* in the source, not year of publication. Publication details can then be found in Appendix 6.

5.3.2 *Area and Main Country Analysis*

5.3.2.1
Tables appearing in each of the main sources for annual data (see section 5.3.1.7) facilitate the compilation of a time series for annual direct investment overseas, by area and country. These tables are consistently numbered (i.e. table 3.1 in [QRL.10] for 1975 to 1984; table 16 or 17 in [QRL.10] for 1968 to 1974; table A in [QRL.5] for 1965 to 1967). The time series that can be constructed from these sources covers the years 1960 to 1984, but excluding insurance for 1960 to 1962. The organisation of area coverage varies in the manner described in section 5.1.5.

5.3.3 *Industry and Country Analysis*

5.3.3.1
Data for UK annual overseas direct investment by industry (of overseas affiliate) and country can be found in the main sources (see 5.3.1.7), providing such coverage for 1965 to 1984. The tables providing this coverage are consistently numbered (i.e. table 3.3 in [QRL.10] for 1975 to 1984; table 18 or 19 in [QRL.10] for 1968 to 1974; table C in [QRL.5] for 1965 to 1967).

5.3.3.2
In the 1984 *Business Monitor* 19 industries are covered (see Appendix 2). As noted earlier (section 5.3.1.6) the introduction of coverage of oil companies in 1984 is as an extra part of the mining sector already covered, not as an additional independent industry. This industry breakdown was that used since 1972, except that until 1976 'shipping' had been separated from 'transportation'. From 1965 to 1971 coverage was restricted to a total of 10 industries, with manufacturing divided into 'electrical and mechanical engineering', 'vehicles and shipbuilding' and 'other manufacturing'. The organisation of area coverage varies in the manner described in section 5.1.5.

5.3.4 *Components of Investment*

5.3.4.1
Accurate interpretation of the causes and motivations underlying annual variations in new UK direct investment overseas is greatly facilitated by knowledge of the components of such investment. It is thus valuable that quite extensive documentation of such components has been a traditional feature of the annual direct investment surveys.

5.3.4.2
In the 1984 coverage ([QRL.10]; 1984 edition) table 1.1 divides annual total net investment into four main components:

(1) Unremitted profits of overseas subsidiaries and associates
(2) Net acquisitions of overseas subsidiaries and associates share and loan capital
(3) Increase in amount due to UK parent companies on inter-company account (with subsidiaries and associates)
(4) Increase in indebtedness to UK parent companies on branch-head office account (for overseas branches).

Unremitted profits is also subdivided into:

(a) UK companies share of overseas companies net profits (profits are after deducting provisions for depreciation and overseas tax on profits and dividends)

(b) dividends received by UK companies (net of any overseas tax)

Then unremitted profits is (a) less (b).

Net acquisitions is also subdivided into:
(c) gross acquisitions of overseas companies share and loan capital
(d) gross disposals of overseas companies share and loan capital.
Then net acquisitions is (c) less (d). Data on this subdivision of net acquisitions is only available from 1975 onwards, and only reported in [QRL.10] from the 1982 edition onwards.

The 1978 *Business Monitor* ([QRL.10]; 1978 edition; table 1.1) and subsequent editions permit the establishment of a time series for this data from 1960 to 1984.

The 1977 coverage ([QRL.10]; 1977 edition; table 1.1) also provides data (for 1960 to 1977) which divided 'provision for depreciation of capital assets' into:
(e) UK companies share of depreciation of overseas subsidiaries and associates
(f) provision for depreciation of capital assets of UK branches overseas.

5.3.4.3

The information described in section 5.3.4.2 above is also presented with a broad area breakdown, i.e. (1) Western Europe, (2) North America, (3) Other Developed Countries, (4) Rest of World. Table 1.3 in the 1976 *Business Monitor M4* and subsequent editions permit the creation of this time series for 1966 to 1984.

5.3.4.4

Table 1.4 in 1984 *Business Monitor MA4* breaks down total net investment into four main components described in 5.3.4.2, the four areas listed in 5.3.4.3, and by main activity (manufacturing; non-manufacturing). Data on this basis are only available from the 1981 *Business Monitor MA4* onwards, this permitting the creation of a time series for 1971 to 1984.

5.3.4.5

More extensive area and country detail on the components of outward investment are given in table 3.5 of 1984 *Business Monitor MA4*. Here information is given on the four main components listed in section 5.3.4.2, and on the breakdown of 'unremitted profits' noted there, but not the breakdown of 'net acquisitions'.

Each of the annual main sources (see section 5.3.1.7) provides similar data on the four main components in detailed country breakdown, but the tables vary in the other information reported. Thus, for example, the information on breakdowns of 'unremitted profits' was only incorporated from the 1976 *Business Monitor MA4* onwards. By contrast up to the 1977 *Business Monitor MA4* information had been provided on (a) 'related trade credit' transactions between UK companies and their overseas affiliates and (b) provision for depreciation of capital assets of overseas affiliates.

The 1981 to 1984 editions of *Business Monitor MA4* provided information (table 3.6) on the four main components (and breakdown of unremitted profits) for 'manufacturing' only, by country. This permits the creation of a time series on this basis covering 1977 to 1984.

5.3.5 *Investment Earnings, by Area and Main Country*

5.3.5.1

Tables appearing in each of the main sources for annual data (see section 5.3.1.7) then permit the compilation of a time series for annual overseas direct investment earnings, by area and country. These tables are mainly consistently numbered (i.e. table 3.2 or 3.5 in [QRL.10] from 1975 to 1984; table 21 or 22 in [QRL.10] from 1968 to 1974; table F in [QRL.5] from 1965 to 1967). The time series that can be constructed from these sources covers the years 1960 to 1984, excluding insurance from 1960 to 1962. The organisation of area coverage varies in the manner described in section 5.1.5.

5.3.6 *Investment Earnings, by Industry and Country*

5.3.6.1

Data for UK annual overseas direct investment earnings by industry (of overseas affiliate) and country can be found in the main sources (see section 5.3.1.7), providing such coverage for 1965 to 1984. The tables providing this coverage are consistently numbered (i.e. table 3.4 or 3.7 in [QRL.10] for 1975 to 1984; table 23 or 24 in [QRL.10] for 1968 to 1974; table H in [QRL.5] for 1965 to 1967).

5.3.6.2

In the 1984 *Business Monitor MA4* 19 industries are covered (see Appendix 2), with oil (first introduced in that year) incorporated as part of the mining sector. This industry breakdown was that used since 1972, except that until 1976 'shipping' had been separated from 'transportation', and that 'shipbuilding' had only been covered as a separate industry since 1975. From 1965 to 1971 coverage was restricted to a total of 10 industries, with manufacturing limited to three broad sectors. The organisation of area coverage varies in the manner described in section 5.1.5.

5.3.7 *Investment Earnings: Components of Earnings*

5.3.7.1

In the 1984 coverage ([QRL.10]; 1984 edition) table 1.1 divides total annual net earnings of overseas direct investment into three components:

(1) UK companies' share of overseas subsidiaries and associates net profits (after deducting provisions for depreciation and overseas tax on profits and dividends)
(2) Interest received by UK companies from overseas subsidiaries and associates (net of any overseas tax)
(3) Net profits of branches overseas (after deducting provision for depreciation and overseas tax on profits and dividends).

The 1978 *Business Monitor MA4* and subsequent editions (table 1.1 in all issues) permit the establishment of a time series for this data from 1960 to 1984, excluding insurance from 1960 to 1962.

5.3.7.2

The information described in section 5.3.7.1 is also presented with a broad area breakdown, i.e. (1) Western Europe, (2) North America, (3) Other Developed Countries, (4) Rest of World. Table 1.3 in the 1976 *Business Monitor MA4*, and the same table in subsequent editions, permit the creation of this time series for 1966 to 1984.

5.3.7.3

More extensive area and country detail on the three components of outward direct investment earnings (section 5.3.7.1 above) are given in table 3.5 of 1984 *Business Monitor MA4* [QRL.10]. Similar data is available in each earlier main source (i.e. table 3.5 or 3.8 in [QRL.10] from 1975 to 1984; table 24 or 25 in [QRL.10] from 1968 to 1974 table I in [QRL.5] from 1965 to 1967).

Breakdowns of outward direct investment earnings by component, by area and country, for 'manufacturing' only can be found in table 3.6 of 1984 *Business Monitor MA4*. This information is available for each *Business Monitor MA4* edition from 1981 onwards, providing a time series covering 1977 to 1984.

5.3.7.4

Up to the 1980 *Business Monitor MA4* ([QRL.10]; 1980 edition; table 3.6) each main source (see section 5.3.1.7) had provided information (by area and country) dividing net earnings of direct investment as (a) profit and interest, minus (b) losses.

5.3.8 *Flows of Payments for Royalties and Services from Affiliates*

5.3.8.1

An overview of the composition of annual net receipts from overseas affiliates has been available from each of the annual main sources (see section 5.3.1.7) on a consistent basis. This covers:

(1) Net profits of branches overseas (after deducting provisions for depreciation and overseas tax)
(2) Dividends received from overseas subsidiaries and associates (net of any overseas tax)
(3) Interest received from overseas subsidiaries and associates (net of any overseas tax)
(4) Royalties receipts from overseas subsidiaries, associates and branches (net of any overseas tax)
(5) Services receipts from overseas subsidiaries, associates and branches (net of any overseas tax)

(6) Remitted profits plus other receipts; sum of (1) to (5)
(7) Royalties payments to overseas subsidiaries, associates and branches (net of any UK witholding tax)
(8) Services payments to overseas subsidiaries, associates and branches (net of any UK witholding tax)

Thus (6) less (7) less (8) measures total remitted profits of overseas affiliates plus other net receipts. This differs from the earnings data covered in 5.3.5 to 5.3.7 by excluding unremitted earnings of overseas affiliates (all branch profits are deemed to be remitted) and by including other transfers (e.g. royalties and services fees). Thus whilst 5.3.5 to 5.3.7 covers the annual earnings of the overseas affiliates from their operations, the data here covers net annual income of the UK parents from their overseas affiliates.

5.3.8.2
In the *Business Monitor MA4/M4* [QRL.10] series the tables providing this information have been consistently numbered; being table 1.5 in the 1976 to 1984 editions, table 1.7 in the 1975 edition, and table 9 in the 1968 to 1974 editions.

5.3.8.3
For all years except 1984 all components listed in 5.3.8.1 exclude oil. For 1984 oil is still excluded from components (4), (5), (7) and (8).

5.3.8.4
The coverage in the 1975 to 1984 *Business Monitor MA4/M4* provides an area breakdown into (1) Western Europe, (2) North America, (3) Other Developed Countries, (4) Rest of World. This permits a time series from 1975 to 1984 on this basis. In the earlier main sources the area breakdown was (1) Overseas Sterling Area; subdivided into developed countries and developing countries; (2) Rest of World; subdivided into developed countries and developing countries. The time series on this basis runs from 1965 to 1974.

5.4 Direct Investment in the UK—Book Values

5.4.1 *Introduction*

5.4.1.1
Inquiries into the book values of foreign direct investments in the UK have been carried out, and reported, simultaneously with the outward investment surveys introduced earlier (section 5.2.1). Like the outward assets data the inward coverage predominantly excludes oil, banking and insurance, with selective coverage of these industries being introduced in recent inquiries. The inquiries on inward investment remain voluntary, with the response considered "fairly good" ([QRL.10]; 1981

Supplement; page 4), although obviously less than that for the now statutory outward investment inquiry.

5.4.1.2

The net book value of an inward direct investment is the value of a foreign company's investment in its UK subsidiaries, branches and associates.

The elements included in book values of direct investment subsidiaries and branches are described in detail in section 5.2.1.2 above. Often, however, a UK subsidiary or branch of an overseas company will have economic links with other overseas affiliates of its parent. In such a case any amounts due to or from other overseas companies belonging to the same group as the overseas parent, are treated as if they were due to or from the investing company itself, and are thus included as part of the net direct investment assets in the UK attributable to the overseas parent company. Borrowing in the UK by foreign-based companies does not itself count as direct investment, but any direct investment assets purchased in the UK from such borrowing is included. However, borrowing in the UK by a UK-based foreign affiliate itself is not included as inward direct investment, since it does not increase the value of net assets attributable to the overseas investing company.

Overseas investment in UK associates consists of the proportion of the UK companies share capital and reserves attributable to the overseas investing company, plus loans and net current balances due to the overseas company. The figures are derived from the accounts of the UK company in which the investment is held.

5.4.2 *Country Data*

5.4.2.1

The census of inward direct investment assets for end-1981 ([QRL.10]; 1981 Supplement) was the first to produce (table 14) a book value for all industries; including oil, banking and insurance. Limited area and country breakdown is provided. The census of end-1978 ([QRL.10]; 1978 Supplement) had been the first to incorporate book value data for oil and banking, leaving only insurance excluded for that year. Some of this data was subsequently revised, and the definitive revised data for 1978 were published in the 1981 *Business Monitor MA4* ([QRL.10]; 1981 Supplement; Table 14A).

5.4.2.2

The basic source for a time series of data on book values of net assets of foreign direct investment in UK (excluding oil, banking and insurance), with area and country detail, is table 15 in the 1981 *Business Monitor MA4* ([QRL.10]; 1981 Supplement). This includes data for each of the years covered by asset inquiries. It should be observed, however, that some of the earlier sources reporting asset data (see Appendix 5) interpolate estimated values for years between inquiries, including for 1960 and 1961, i.e. prior to the first survey ([QRL.5]; 26th January 1968; table 17).

5.4.3 *Industry Data*

5.4.3.1

A time series of data dividing foreign direct investment assets in the UK into 'manufacturing' and 'non-manufacturing' appears in the 1981 *Business Monitor* MA4 ([QRL.10]; 1981 Supplement; table 17). This covers all the years of book value inquiries, other than 1962. The area coverage of investing countries comprises:
(1) Western Europe; subdivided into EC and EFTA
(2) North America; subdivided into Canada and USA
(3) Other Developed Countries
(4) Rest of World.

5.4.3.2

With the exception of that for 1962, each of the inward direct investment asset surveys has produced a matrix table of data for book values by investing country and by industry of UK affiliate. The table for 1981 ([QRL.10]; 1981 Supplement; table 19) covers 16 industries in total. Of these 10 are manufacturing industries, with 'shipbuilding' not covered separately by comparison with the outward asset coverage. The six non-manufacturing industries included in the inward asset table for 1981 differ substantially from the eight in the comparable outward table (see section 5.2.3.2), adding 'shipping' but excluding 'agriculture, forestry and fishing', 'mining and quarrying' and 'property owning and managing'.

The industry coverage for inward direct investment assets was the same as for 1981 in 1978, and also in 1974 except that 'property owning and managing' was given separate coverage. The industry coverage in 1965, 1968 and 1971 was the same as for the later years with respect to manufacturing, but non-manufacturing was only divided into 'distribution' and 'other activities'. The area/country coverage in these tables was very broad in the earlier years, but more detail has been provided in the more recent inquiries. It should be noted that the 1968 data is available from the *Board of Trade Journal* source ([QRL.5]; 23rd September 1970) but not the equivalent *Business Monitor M4*.

5.4.4 *Forms of Organisation*

5.4.4.1

Data from the 1981 inquiry ([QRL.10]; 1981 Supplement; table 18) provide information, with a quite detailed area and country breakdown, on the organisational form taken by foreign direct investment in the UK. Thus investment is divided into:
(1) Subsidiaries: (a) number, (b) book value
(2) Branches: (a) number, (b) book value
(3) Associates: (a) number, (b) book value
(4) Total: (a) number, (b) book value

The 1974 and 1978 inquiries provided data on the same basis.

5.4.4.2

Similar data to that for 1974-81 were derived in the 1965, 1968 and 1971 inquiries. Here investment is divided into:

(1) Large investing companies—subsidiaries (a) number (b) book value
(2) Large investing companies—branches (a) number (b) book value
(3) Large investing companies—associates (a) number (b) book value
(4) Small investing companies (a) number (b) book value

The limit for small investing companies was £1 million in 1971, £0.5 million in 1968, and £0.25 million in 1965. The data for 1968 are available from the *Board of Trade Journal* ([QRL.5]; 23rd September 1970; table 5), but not from the equivalent *Business Monitor M4*.

5.4.5 *Percentage Shareholding*

5.4.5.1

Data from the 1981 inquiry ([QRL.10]; 1981 Supplement; table 20) provide information, with area and country breakdown, on overseas investors percentage shareholding in foreign direct investment subsidiaries in UK. The ownership categories covered are the same as those listed for outward investment earlier (see 5.2.5.1). The 1974 and 1978 inquiries provided similar data.

5.4.5.2

The 1981 inquiry ([QRL.10]; 1981 Supplement; table 21) also performed an analysis of ownership patterns of UK associates of foreign investors, comparable to that for subsidiaries described above. The ownership categories are the same as those covered by the comparable analysis for UK overseas associates (see 5.2.5.2). The 1974 and 1978 inquiries provided similar data.

5.4.5.3

Since branches are, by definition, 100 per cent owned, these are not covered in this way.

5.4.5.4

Some information on ownership patterns of inward direct investment was derived from the 1965 inquiry. Thus table 24 ([QRL.5]; 26th January 1968) provides information in the categories of section 5.4.5.1 above, but for subsidiaries and associates combined. Both number of affiliates and book values are covered, but investing country coverage is limited to USA/Rest of World. Table 25 provides similar information on ownership patterns with an industry breakdown.

5.4.6 *Size of Enterprise Data*

5.4.6.1
Some indication of the size distribution of foreign direct investment assets in the UK has been provided by recent inquiries. Thus in the report of the 1981 inquiry ([QRL.10]; 1981 Supplement) table 24 divides UK enterprises into nine size groups ('under £1 million' to 'over £150 million') according to net assets attributable to overseas investors. For each size group data are given on:
(1) Number of UK enterprises
(2) Number of UK subsidiaries, branches and associates
(3) Book values of net assets of UK affiliates.
Similar data are available for 1978, for 1974 (for only eight size groups) and for 1971 (for only seven size groups).

5.4.7 *Balance Sheet Data of Foreign Subsidiaries in the UK*

5.4.7.1
As was the case for outward direct investment each of the inquiries into foreign direct investment in the UK has provided information, with varying degrees of coverage by investing area and country, on the balance sheet structure of foreign subsidiaries in the UK. It should be noted, however, that the first (i.e. 1962) inquiry ([QRL.5]; 7th August 1964; table 4) did not separate data for subsidiaries and branches.

5.4.7.2
Apart from obvious adjustments in terminology to allow for the reversed direction of ownership, the balance sheet analysis for foreign subsidiaries in the UK for end-1981 ([QRL.10]; 1981 Supplement; table 22) is organised precisely as that for outward subsidiaries described above (see section 5.2.7.2).

5.4.7.3
The balance sheet analysis of inward subsidiaries for 1978 ([QRL.10]; 1978 Supplement; table 21) and 1974 ([QRL.10]; 1974 Supplement; table 18) differs from that for 1981 only in the minor ways reported for outward investment in sections 5.2.7.3 and 5.2.7.4 above.

5.4.7.4
Again apart from the obvious differences in terminology to allow for the reversed direction of ownership, the balance sheet analysis for foreign subsidiaries in the UK for end-1971 ([QRL.10]; 1972 edition part 11; table 46) is organised in the same way as that for outward subsidiaries described in section 5.2.7.5 above. The data for 1971

are reported as 'returns received' in the inquiry, without the estimated allowance for non-response incorporated in later years. The coverage for 1968 ([QRL.10]; 1969 edition; table 39) is as for 1971, and again limited to 'returns received' in the inquiry. Apart from one difference (described in the comparable outward investment section—see section 5.2.7.7) the 1965 coverage ([QRL.5]; 26th January 1968; table 21) is organised as for 1968 and 1971, but differs from those years by allowing for non-response but only covering large investments.

5.4.8 *Balance Sheet Data of Foreign Branches in the UK*

5.4.8.1

The coverage of the balance sheet analysis for foreign branches in the UK at end-1981 ([QRL.10]; 1981 Supplement; table 23) is the same as for outward branches (section 5.2.8.2), except that branches' own 'investment in subsidiaries' is an additional component of net assets. The coverage for 1978 ([QRL.10]; 1978 Supplement; table 22) differs from that for 1981 only in the way noted in section 5.2.8.3; whilst that for 1974 ([QRL.10]; 1974 Supplement; table 17) differs from that for 1978 as reported in section 5.2.8.4.

5.4.8.2

The organisation of inward branch balance sheet coverage for 1968 and 1971 (in both years data covering only 'returns received') is as for outward branches described in section 5.2.8.5; except (i) that 'current assets' excludes amount due to overseas parent, and (ii) that 'current liabilities' excludes amount due to overseas parents.

5.4.8.3

The data for 1965 ([QRL.5]; 26th January 1968; table 22) provide allowance for non-response, but are restricted to 'large investments'. Otherwise the coverage is as for 1968 and 1971, except that 'long-term liabilities in the UK and provision for future taxation' are not separated as later.

5.4.9 *Date of Establishment Data (1965 Inquiry)*

5.4.9.1

As was reported in the discussion of UK outward direct investment assets (see section 5.2.9), the 1965 survey collected information on the date of establishment of the operations covered. Similar information is provided for inward direct investment ([QRL.5]; 26th January 1968). Once again the analysis is restricted to large investing companies and covers the time periods:

(1) before 1946
(2) 1946-55
(3) 1956-65

5.4.9.2
In table 26 the end-1965 assets and numbers of affiliates in UK are divided by industry (twelve industries, of which ten are manufacturing) and by period of establishment. Information on form of organisation (branch; subsidiary; associate) and period of establishment is given in table 23, which also divides investments by source (USA; Rest of World). Two smaller tables, in the text of the *Board of Trade Journal* article, relate date of establishment of the foreign affiliate in the UK to (a) size of investment at end-1965, and (b) degree of foreign ownership of the investment at end-1965.

5.5 Foreign Direct Investment in the UK—Annual Transactions

5.5.1 *Introduction and Definitions*

5.5.1.1
The transactions covered by the annual surveys may be divided into those relating to new foreign direct investment in the UK (see sections 5.5.2 to 5.5.4) and those relating to the earnings on foreign direct investment in the UK (see sections 5.5.5 to 5.5.8).

5.5.1.2
For inward direct investment the statistics cover all UK companies in which an overseas resident has a direct shareholding of twenty per cent or more, subject to the same exceptions as noted (see section 5.3.1.2) for outward direct investment, plus all subsidiaries and associates of these companies, plus all unincorporated businesses and branches in the UK owned by overseas residents (see Walker; [B.24]; p 61.2).

5.5.1.3
The components of annual inward direct investment are as described in section 5.3.1.3 for outward direct investment.

5.5.1.4
The components of inward direct investment earnings are as described in section 5.3.1.4 for outward direct investment. It should be observed (see Walker; [B.24]; p 61.2) that in the UK coverage earnings due abroad on inward direct investment include those of any direct investment enterprise in a third country controlled by a foreign direct investment enterprise in the UK.

5.5.1.5
Annual foreign direct investment in the UK is normally recorded *net* in the tables reviewed below, i.e. the figures represent the difference between gross direct

investment in the UK and gross disinvestment from the UK. Up to the 1981 *Business Monitor MA4* ([QRL.10]; 1981 edition; table 2.6) data had been given on gross investment and gross disinvestment, including analyses by components.

5.5.1.6
Oil companies are included in the coverage of inward direct investment for the first time in 1984, though only as a part of the 'other non-manufacturing activities' sector in the detailed industry breakdown. Coverage of 'insurance' only commenced in the 1978 *Business Monitor MA4* [QRL.10]; though some of the time series in that edition permit retrospective incorporation of insurance for some earlier years. The details of this are given as appropriate below.

5.5.1.7
The 'main sources' distinguished for data on inward direct investment are as for outward (see section 5.3.1.7 and also section 5.3.1.8).

5.5.2 *Area and Country Analysis*

5.5.2.1
Tables appearing in each of the main sources for annual data (see section 5.3.1.7) facilitate the compilation of a time series of annual foreign direct investment in the UK, by area and country of investor. These tables are consistently numbered (i.e. table 4.1 in [QRL.10] for 1975 to 1984; table 26 or 27 in [QRL.10] for 1968 to 1974; table K in [QRL.5] for 1965 to 1967). It should be noted that the data reported in the sources up to the 1977 edition of *Business Monitor MA4* exclude oil and insurance, whilst those from 1978 to 1983 exclude only oil. With allowance for the treatment of insurance a time series can be constructed from the sources for 1960 to 1984. The organisation of area coverage varies in the manner described in section 5.1.5.

5.5.3 *Industry and Country Analysis*

5.5.3.1
Data for annual foreign direct investment in the UK by industry (of UK affiliate) and investing country can be found in the main sources (see section 5.3.1.7), providing such coverage for 1965 to 1984. The tables providing this coverage are consistently numbered (i.e. table 4.3 in [QRL.10] for 1975 to 1984; table 28 or 29 in [QRL.10] for 1968 to 1974; table M in [QRL.5] for 1965 to 1967). The incorporation of coverage of insurance is as noted in section 5.5.2.1.

5.5.3.2
In the 1984 *Business Monitor MA4* fifteeen industries are covered. By comparison with the corresponding outward investment table (see Appendix 2), 'shipbuilding' is omitted as a separate industry and included in 'other manufacturing', whilst 'agriculture', 'mining' and 'transport' are also omitted as separate industries and included in 'other non-manufacturing'. This industry breakdown was that used since 1975, except that in 1975 and 1976 'shipping' was treated as a separate industry, rather than as a component of 'other non-manufacturing' as in subsequent years. From 1972 to 1974 the coverage used ten manufacturing industries (as 1975 onwards), but divided non-manufacturing into only 'distribution' and 'other activities'. From 1965 to 1971 only four industries were covered, with non-manufacturing divided as just noted for 1972 to 1974, and manufacturing separated into 'electrical and mechanical engineering' and 'other'. The organisation of area coverage varies as described in section 5.1.5.

5.5.4 *Components of Investment*

5.5.4.1
As was observed in discussing outward investment (see section 5.3.4.1) analysis of direct investment flows is greatly sharpened by knowledge of the behaviour of the various components of that investment. Below we survey the quite extensive data provided on these components for foreign direct investment in the UK.

5.5.4.2
In the 1984 coverage ([QRL.10]; 1984 edition) table 2.1 divides annual total net investment into four main components. With due allowance in terminology for the reverse direction of transaction these components are those described for outward investment in section 5.3.4.2.

The further subdivision of two of these components, 'unremitted profits' and 'net acquisitions', is also available for inward investment in a form equivalent to that for outward investment described in section 5.3.4.2.

The 1978 *Business Monitor MA4* ([QRL.10]; 1978 edition; table 2.1) and subsequent editions permit the establishment of a time series for this data from 1960 to 1984. This time series omits insurance prior to 1969, although inward insurance investments are believed to be small prior to 1969, so significant distortion is unlikely from this change in coverage.

The 1977 coverage ([QRL.10]; 1977 edition; table 2.1) also provides data (excluding insurance throughout) on 'provision for depreciation' for 1960 to 1977. The subdivision of this is again analogous to that delineated in 5.3.4.2.

5.5.4.3
The information described in section 5.3.4.2 above is also presented with a broad area breakdown of investing country, i.e. (1) Western Europe, (2) North America, (3)

Other Developed Countries, (4) Rest of World. Table 2.3 in the 1976 *Business Monitor M4* and subsequent editions permit the creation of this time series for 1966 to 1984, although insurance companies could not be incorporated before 1969.

5.5.4.4

Table 2.4 in the 1984 *Business Monitor MA4* breaks down total net investment into *three* main components (combining the last two of the four originally listed in section 5.3.4.2), for the four source areas listed above (section 5.5.4.3) and by main activity (manufacturing; non-manufacturing). Data on this basis are available from the 1981 *Business Monitor MA4* onwards, permitting a time series from 1971 to 1984.

5.5.4.5

More extensive investing area and country detail on the components of inward direct investment are given in table 4.5 of 1984 *Business Monitor MA4*. Here information is given on three main components (see section 5.5.4.4 above), and a subdivision of 'unremitted profits'.

Each of the annual main sources (see section 5.3.1.7) provides similar data on the three main components in detailed country breakdowns, but the tables vary in the other information reported. Thus data on the subdivision of 'unremitted profits' is only available from the 1976 *Business Monitor M4* onwards, whilst coverage of 'provisions for depreciation' are available only up to the 1977 *Business Monitor MA4*.

The 1981 to 1984 editions of *Business Monitor MA4* provided information (table 4.6) on the three main components (see section 5.5.4.4) of investment, and the further division of 'unremitted profits', for manufacturing industry only. A time series on this basis is thus available for 1977 to 1984.

5.5.5 *Earnings Data by Investing Area and Country*

5.5.5.1

Tables appearing in each of the main sources for annual data (see section 5.3.1.7) permit the construction of a time series of annual earnings of foreign direct investment in the UK, by investing area and country. These tables are consistently numbered (i.e. table 4.2 or 4.5 in [QRL.10] for 1975 to 1984; table 31 or 32 in [QRL.10] for 1968 to 1974; table P in [QRL.5] for 1965 to 1967). Sources up to the 1977 edition of *Business Monitor MA4* exclude insurance. With allowance for the coverage of insurance a time series can be constructed from 1960 to 1984 from these sources. The organisation of area coverage varies in the manner described in section 5.1.5.

5.5.6 *Earnings Data by Industry and Investing Country*

5.5.6.1
Data on earnings of foreign direct investment in the UK, by industry of UK affiliate and investing country, is only provided with detailed coverage in the *Business Monitor M4/MA4* from 1973 to 1984. The tables are consistently numbered (i.e. table 4.4 or 4.7 in [QRL.10] for 1975 to 1984; table 34 in [QRL.10] for 1973 and 1974). Insurance is only included in the coverage for 1978 *Business Monitor M4* onwards.

5.5.6.2
The industry coverage for these tables is as for the period 1973 to 1984 for the inward investment data described in section 5.5.3.2.

5.5.7 *Earnings Data by Components of Earnings*

5.5.7.1
In the 1984 coverage ([QRL.10]; 1984 edition) table 2.1 divides total annual net earnings of foreign direct investment in UK into three components. With allowance for the reversed direction of transaction these components are as listed in section 5.3.7.1. The 1978 *Business Monitor MA4* and subsequent editions (table 2.1 in all issues) permit the establishment of a time series for this data from 1960 to 1984. This time series excludes insurance prior to 1969, although it seems unlikely that insurance companies transactions before 1969 were of sufficient magnitude for their omission to significantly distort the results.

5.5.7.2
The information described in section 5.5.7.1 above is also presented broken down by area of investing country. The areas delineated are (1) Western Europe, (2) North America, (3) Other Developed Countries, (4) Rest of World. Table 2.3 in the 1976 *Business Monitor M4*, and the same table in subsequent editions, permit the creation of time series for 1966 to 1984. Insurance cannot be included in this series earlier than 1969.

5.5.7.3
More extensive investing area and country detail on the three components of foreign direct investment earnings in UK are given in table 4.5 of the 1984 *Business Monitor MA4* [QRL.10]. Similar data is available in *Business Monitor M4/MA4* from 1973 to 1984 (table 4.5 in 1976 to 1984; table 3.5 in 1973 to 1974; 1975 data covered by later year's tables).

A breakdown of inward direct investment by component, and by area and country, for manufacturing industry only, can be found in table 4.6 in each *Business Monitor MA4* from 1981 onwards, providing a time series covering 1977 to 1984.

5.5.7.4
Up to the 1980 *Business Monitor MA4* ([QRL.10]; 1980 edition; table 4.6) each main source (see section 5.3.1.7) had provided information (by investing area and country) dividing net earnings of inward direct investment as (a) profit and interest, minus (b) losses.

5.5.8 *Earnings from UK Affiliates to Overseas Investors*

5.5.8.1
An overview of the composition of annual net payments by UK affiliates to their overseas parents and associated companies has been available from each of the main sources (see section 5.3.1.7). The eight components of these data are, with due allowance in terminology for the reversed direction of transactions, those listed in section 5.3.8.1 for outward investment.

The aggregate for these components measures total profits remitted by the foreign affiliates in the UK plus their other net payments, and thus differs from the earnings covered in sections 5.5.5 to 5.5.7 by excluding unremitted profits but including other transfers (e.g. royalties and service fees).

5.5.8.2
In the *Business Monitor M4/MA4* [QRL.10] series, tables embodying this information have been consistently numbered; being table 2.5 or 2.6 from 1975 to 1984 and table 15 or 16 from 1968 to 1974.

5.5.8.3
For all years except 1984 all components exclude oil. For 1984 oil is still excluded from components (4), (5), (7) and (8) (see section 5.3.8.1). All sources up to and including 1977 *Business Monitor MA4* omit coverage of insurance, although use of the 1978 *Business Monitor MA4* would permit inclusion of insurance in a time series from 1975 onwards.

5.5.8.4
The coverage in the 1975 to 1984 *Business Monitor M4/MA4* provides an area division into (1) Western Europe, (2) North America, (3) Other Developed Countries, (4) Rest of World. From 1966 to 1974 the coverage is restricted to USA/Other Countries; though more detail was provided in 1965.

5.6 Foreign Enterprises in UK Manufacturing Industry—Census of Production Data

5.6.1 *Introduction*

5.6.1.1
Information on the activity of foreign enterprises in UK manufacturing industry has been collected in the Census of Production since 1963. In the annual series of Censuses undertaken since 1971 detailed coverage of foreign enterprises was provided biennially from 1973 to 1983 but with 1984 also covered. Information on the participation of foreign enterprises amongst the largest 100 UK enterprises is provided annually since 1973.

5.6.1.2
The Census of Production data broaden the perspective on inward direct investment in the UK beyond that covered by the sources reviewed in section 5.4 and 5.5, by covering several variables (e.g. employment; sales; wages and salaries) not covered there. Also direct and precise figures for degree of participation of foreign enterprises in UK manufacturing can be calculated. Similarly, comparative performance measures (e.g. productivity; wage levels etc) for foreign enterprises and all UK manufacturing industry are immediately accessible.

5.6.1.3
The information for 1973 to 1984 is contained in each year's Summary Volume of the Report of the Census of Production, published as *Business Monitor* PA1002 [QRL.17]. Information for 1968 is published in the summary tables to that Census [QRL.52], and information for 1963 in that year's summary tables [QRL.53].

5.6.2 *Country of Origin Data*

5.6.2.1
In the 1984 *Census of Production Summary Volume* [QRL.17] Table 18 gives data on twenty measures of participation, activity or performance in 1984. These are:
(1) Number of Enterprises
(2) Number of Establishments
(3) Total employment; i.e. (4)+(5)
(4) Total operatives
(5) Total other employees
(6) Total wages and salaries of operatives
(7) Wages and salaries of operatives per head
(8) Total wages and salaries of other employees
(9) Wages and salaries of other employees per head
(10) Total sales and work done
(11) Gross output

(12) Total net output
(13) Net output per head
(14) Total gross value added at factor cost
(15) Total gross value added at factor cost per head
(16) Total net capital expenditure; i.e. (17) less (18)
(17) Acquisitions
(18) Disposals
(19) Total stocks and work in progress—increase during year
(20) Total stocks and work in progress—value at year end

5.6.2.2
The range of information listed in section 5.6.2.1 is given by country of origin of enterprise, this being grouped as:
(1) Total of all enterprises in UK classified to manufacturing
(2) Total of all foreign enterprises in UK classified to manufacturing
(3) Total European Community enterprises in UK classified to manufacturing, with data on seven individual countries
(4) Total of all other foreign enterprises in UK classified to manufacturing, with data on ten individual countries

5.6.2.3
This information is available (with some variation in countries covered) on a biennial basis, starting in 1973 but with 1984 also covered. In the Census of Production Summary Volumes [QRL.17] this data is at Table 18 in all years, except 1979 (table 19).

5.6.2.4
The 1968 *Census of Production Summary Volume* ([QRL.52]; table 43) provides data similar to that outlined above. Seven measures of participation or performance are covered, with data provided for UK firms and for those of eleven foreign nationalities. Data for 1963, similar to that for 1968, is given in table 20 of the *Census of Production Summary Volume* for 1963 [QRL.53], though it should be noted that the originally published results in this table were subject to subsequent revision.

5.6.2.5
Similar data to that discussed above were produced in the 1971 Census ([QRL.17]; 1971 edition; table 14). The figures are, however, only for 'returns received' for overseas controlled enterprises, and include 'mining and quarrying' in addition to manufacturing.

5.6.3 *Industry Data*

5.6.3.1
In the 1984 *Census of Production Summary Volume* [QRL.17]; table 19 gives data on nineteen measures of participation, activity or performance in 1984. These are the twenty measures listed in section 5.6.2.1, excluding 'number of enterprises'.

5.6.3.2
The information is given subdivided by manufacturing industry. In the 1981 to 1984 Censuses twenty-one industries are listed (based on the 1980 SIC classes), though some data are suppressed for reasons of disclosure and confidentiality. From 1973 to 1977 fifteen industries were covered, based on 1968 SIC Orders, with this expanded to 16 in 1979.

5.6.3.3
The information is available on a biennial basis starting in 1973 but with 1984 also covered. In the Census Summary Volumes [QRL.17] these data are at table 19 in all years, except 1979 (table 20).

5.6.3.4
The 1968 *Census of Production Summary Volume* ([QRL.52]; table 44) provides data, for 1963 and 1968, on 'sales of foreign owned enterprises as a percentage of total sales' by 1968 SIC minimum list heading.

5.6.3.5
The 1971 *Census of Production* produced data similar to that discussed above ([QRL.17]; 1971 edition; table 15). The data are, however, limited to 'returns received' and include 'mining and quarrying' in addition to manufacturing.

5.6.4 *Regional Data*

5.6.4.1
In the 1984 *Census of Production Summary Volume* [QRL.17] table 20 gives data, by UK region, on foreign enterprises in UK manufacturing. The measures covered are:
(1) Total employment
(2) Net capital expenditure
(3) Net output
(4) Gross value added at factor cost

5.6.4.2
In addition to data for England, Wales, Scotland and Northern Ireland, eight English regions are covered. The data are also divided, within regions, by type of assisted area. These are (1) non-assisted, (2) special development, (3) development, (4) intermediate and (5) derelict land clearance.

5.6.4.3
This information is available on a biennial basis, starting in 1973 but with 1984 also covered. It should be noted that measures (3) and (4) listed in section 5.6.4.1 are only available in the 1979 coverage onwards. In the Census Summary Volumes [QRL.17] these data are at table 20 in all years, except 1979 (table 21).

5.6.4.4
Table 16 in the *Report of the 1971 Census* [QRL.17] produces similar data for that year, but only on a 'returns received' basis.

5.6.5 *Participation of Foreign Enterprises in UK Private Manufacturing*

5.6.5.1
Interesting annual Census of Production data provide information on the extent of foreign-owned enterprises' participation among the largest firms in the UK manufacturing sector. Thus three tables in the 1984 *Census of Production Summary Volume* [QRL.17] analyse the 100 largest private sector enterprises classified to manufacturing, according to different size criteria.
(i) Table 15 covers the 100 largest enterprises defined by 'size of total employment'
(ii) Table 16 covers the 100 largest enterprises defined by 'size of net output'
(iii) Table 17 covers the 100 largest enterprises defined by 'size of total sales and work done'.

5.6.5.2
In each of the three tables introduced in section 5.6.5.1 data are given on the largest 100 enterprises, total:
(1) Number of Enterprises (not available in years before 1984)
(2) Number of Establishments
(3) Employment
(4) Sales and Work Done
(5) Net Output
(6) Gross Value Added at Factor Cost
(7) Net Capital Expenditure
(8) Stocks and Work in Progress—Increase During Year
(9) Stocks and Work in Progress—Value at End of Year
The contribution of foreign enterprises to each of these eight measures is given in

each table. From 1973 to 1982 footnotes to the tables record the number of foreign enterprises in the top 100 covered.

5.6.5.3
As noted above, this information is available annually from 1973 to 1984. Except for 1979 the tables are numbered as given in section 5.6.5.1.

5.6.6 *Foreign-Owned Manufacturing Units in the UK*

5.6.6.1
The Census of Production data on employment (along with information collected in other Business Statistics Office production inquiries) is incorporated in the BSO register of UK manufacturing units. An analysis of these data is provided in the annual publication *Business Monitor* PA1003—Analysis of United Kingdom Manufacturing (Local) Units by Employment Size [QRL.18]. A local manufacturing unit is a factory or plant located at a single site or address. This is a level of reporting different from that used in the Census of Production reports.

5.6.6.2
In the 1984 edition of this source [QRL.18] four of the tables provide information on foreign-owned manufacturing units. Country of ownership data is covered in Table 16 ([QRL.18]; 1984 edition). Data on total foreign-owned manufacturing units (with 20 or more employees) is given, with information on their employment. Nine individual foreign countries of ownership are covered. Data on numbers of foreign-owned manufacturing units and their employment, by SIC class are given in Table 17 ([QRL.18]; 1984 edition); whilst similar information, broken down by UK country and standard region, are provided in Table 18. Regional data on number of foreign units and their employment are also given in Table 19, this time with a subdivision by employment size group.

5.7 UK Direct Investment—Comparative Position

5.7.1 *Problems of Inter-Country Comparisons*
Although direct investment plays a major role in the contemporary international economy the compilation and publication of centrally coordinated sets of reliable data for all investing and host countries has proved extremely difficult. Thus, although the UK's own direct investment transactions are well covered in the sources reviewed above, comparative aspects of the UK's position in international direct investment are less easy to document.

5.7.2 *IMF Data*

Several tables in Part 2 of the *IMF Balance of Payments Statistics Yearbook* [QRL.3] compile country data on various components of direct investment transactions. Thus tables C-6 and C-7 ([QRL.3]; Part 2; 1986 edition) provide data by country and region on, respectively:

(1) reinvested earnings on direct investment
(2) other direct investment income

However the number of gaps, due to non-availability of data, in these tables (most notably that on reinvested earnings, where many leading investing or host countries do not provide information) constrain the viability of successful use of these tables for inter-country comparisons.

Similarly tables C-13, C-14 and C-15 ([QRL.3]; Part 2; 1986 edition) provide data by country and region on respectively:

(1) net annual direct investment
(2) annual direct investment abroad
(3) annual direct investment in the reporting economy

Again, however, the deficiency in reporting reinvested earnings bedevils these tables. This should be a component of annual investment flows, but is omitted from the coverage of many countries included in the tables.

5.7.3 *IRM Directory*

5.7.3.1

As a leading outward direct investor, and also an important host to the direct investments of other countries, information on UK direct investment often appears in the data sources of other countries. An extremely valuable compilation of such data is contained in the *IRM Directory of Statistics of International Investment and Production* (Dunning and Cantwell; [QRL.40]). This directory incorporates surveys of available direct investment data for eighty countries; these countries being estimated to account for at least ninety-nine per cent of the outward direct investment of companies throughout the world, and between ninety-six per cent and ninety-eight per cent of inward direct investment stake.

5.7.3.2

To the extent permitted by the reporting countries' data availability, an individual country's entry in the Dunning and Cantwell *Directory* [QRL.40] may provide information (table A7 in each country's entry) on (i) the UK share of the country's inward and/or outward direct investment stake in one or more recent years, or (ii) the UK share of the country's cumulative inward and/or outward direct investment flows over a recent period. Each entry also provides a guide to the leading national data sources, from which more information on the UK's direct investment relations with the country may be derived.

5.7.3.3
The Dunning and Cantwell *Directory* [QRL.40] also draws up a number of comparative tables, deriving aggregates from their eighty individual country entries. These tables permit the drawing of, at least tentative, conclusions about the role of the UK in the global direct investment situation, or the comparison of certain characteristics of UK direct investment (inward or outward) with that of other countries.

Thus tables B1, B15 and B16 provide data on UK share of total inward and outward direct investment capital stock, for 1975 and 1983. Tables B2 (for 1975) and B3 (for 1982) divide countries' inward and outward capital stock by sector (primary; secondary; tertiary), thus allowing comparison of the UK with other countries with respect to this characteristic. In a similar way tables B4 (for 1975) and B5 (for 1982) divide countries' inward and outward capital stock between developed and developing countries, again providing a basis for comparison of UK with other countries.

5.8 UK Outward and Inward Direct Investment—Acquisitions

5.8.1 *Introduction*

The quarterly publication *Business Monitor* MQ7—'Acquisitions and Mergers by Industrial and Commercial Companies' [QRL.14] provides data on 'international acquisitions' i.e. takeovers of foreign companies by UK companies and *vice versa*.The data are compiled by the DTI using reports in the financial press, backed by special inquiries to determine the form, value and timing of particular acquisitions. Insurance, banking and other financial companies are not covered.

5.8.2 *Outward Acquisitions*

The acquisition of foreign companies by UK companies is covered by Table 4 in *Business Monitor* MQ7 ([QRL.14]; third quarter 1987). Data is given on:

(1) Number of acquiring companies
(2) Number of companies acquired
(3) Value of acquisitions (£m.)

In addition to total figures, data are given for EEC countries and Other Countries. Annual figures are given from 1969, and quarterly figures for the most recent five years.

The publication notes that this coverage of acquisitions by UK companies of foreign companies is likely to be limited for two reasons. Firstly, press reports of foreign acquisitions are likely to be less complete than those relating to acquisitions within the UK. Secondly, only direct acquisitions are counted, with acquisitions by existing UK foreign subsidiaries or associates ignored. Since many acquisitions of foreign companies are achieved through existing affiliates, the reported figures record only part of the expansion of UK companies abroad.

5.8.3 *Inward Acquisitions*

The acquisition of UK companies by foreign companies is covered by Table 5 ([QRL.14]; third quarter 1987). The format of the data is the same as that described in section 5.8.2 for Table 4. The figures cover only direct acquisitions by foreign registered companies, thus excluding cases where a UK company, which is a subsidiary or associate of a foreign company, acquires another UK company.

CHAPTER 6

ROYALTIES AND MISCELLANEOUS PAYMENTS AND RECEIPTS

6.1 Introduction

In this chapter we review three series of data on overseas transactions, each deriving, at least in part, from surveys carried out by the Department of Trade and Industry and its predecessors.

In section 6.2 we look at data on royalties and similar transactions collected by the DTI as part of its main annual overseas transactions survey.

The royalties data reviewed in section 6.2 do not include the overseas transactions of UK located film and TV companies. Information on this is compiled by the DTI in a separate survey. Section 6.3 outlines the published results from these surveys.

Up to 1981 an important part of the main overseas transactions survey related to UK direct exports and this is considered in section 6.4. In some of the tables reported from these surveys the results were augmented by the use of published (Company Account) data for oil companies (not approached in the survey) and for other large companies which did not respond to the survey. As is explained in section 6.4.1.1 it was not possible to continue this series adequately beyond 1981. The distinguishing feature of the coverage of direct exports in the survey referred to above is the opportunity it provides of analysing exports by category of firm. One use of this is to categorise firms by the size of their annual exports, thus permitting the derivation of evidence on degrees of concentration in exporting. Another important facet of the categorisation of firms is in terms of the nature of their relationship with other firms located overseas. For some years this perspective is further elaborated by the provision of data on 'related' exports, i.e. exports to 'related' enterprises abroad.

6.2 Royalties and Similar Transactions

6.2.1 *Introduction and Definitions*

6.2.1.1

The data on royalties covered by the overseas transactions survey covers amounts payable to, or receivable from, overseas concerns (either 'related' or 'unrelated': see section 6.2.1.3) on current or capital account, after deduction wherever appropriate of UK or overseas tax.

Two types of royalty are distinguished in the data:

(1) artistic royalties; i.e. those on printed matter, sound recordings or performing rights (but excluding film or TV rights)

(2) technological and mineral royalties; including patents, licences, trade marks, designs, copyrights, mining and quarrying mineral rights, manufacturing rights, use of technical 'know-how' and technical assistance.

The figures include amounts paid to agents for services rendered in connection with royalties etc. and also agency earnings of UK concerns and lump sum settlements. However, sales commission and registration fees on patents held by UK concerns are excluded. Other transactions that involve payments for knowledge and similar intangible assets are excluded from the royalties data, if they cannot be adequately separated from other transactions. Thus some royalties transactions are included in the recorded value of UK imports and exports, especially where charges for royalties etc. are incorporated in the contract price of capital goods and cannot be adequately distinguished. A notable example of this is likely to occur in the case of 'turnkey' agreements, where large contracts are made by individual firms or consortia for the erection of complete plants or facilities, and where the transfer of knowledge of how to operate the units is an integral part of the contract. The data also omits any royalties etc. which are included indistinguishably in profits earned by overseas affiliates.

6.2.1.2

It is important to note that the financial receipts and payments relating to royalties and similar transactions are often used as an *indicator* of magnitudes of technology flows. Whereas the scarcity of other indicators for such knowledge flows does impart a value to royalty data in this way, their limitations should be admitted. We have already seen in section 6.2.1.1 that transactions that clearly represent payments for knowledge may not be separable from visible trade, or from profits. Further the market for knowledge is notably imperfect, so that the price determined for technology (even 'arms length' bargains between 'unrelated' parties: see section 6.2.1.3) may not adequately reflect the 'true' value of the knowledge transferred. Thus different technology bargains may embody various degrees of monopoly rent, according to the knowledge or bargaining power of the parties.

6.2.1.3

A valuable facet of the reporting of royalties has been the incorporation of a distinction between transactions involving 'related' concerns, and those between 'unrelated' concerns.

'Related' concerns include overseas parent companies, overseas branches, overseas subsidiaries and overseas associates of the reporting concern. Since 1979 'related' concerns have also included overseas fellow subsidiaries i.e. overseas concerns which have the same ultimate overseas parent company as foreign-owned UK companies and branches, but which themselves have no direct investment in the UK concern.

As is well acknowledged, the price which two related parties decide upon in recording a transfer of technology between them is subject to a wide-range of influences, and may well differ substantially from that which would have been agreed if the parties were bargaining as independent units. Bearing in mind that even 'arms length' technology prices are subject to distortion (see section 6.2.1.2), we should again note in this case that financial flows deriving from technology transfers may only be a very approximate indicator of the intrinsic value of the knowledge involved.

6.2.1.4

An important constraint to be noted with respect to the published royalties data is that the results reported with a country/area breakdown (see section 6.2.3) or an industry breakdown (see section 6.2.4) are derived from 'returns received' in each survey, with no attempt to derive estimates for non-response. In particular, this precludes the calculation of precise net figures for any year, or the derivation of precise indicators of trends in receipts and payments by country or industry. However, the response rate (eighty per cent for receipts, and 90 per cent for payments, is quoted for recent years) seems to be high enough to permit the derivation of a broad picture with respect to both deficit or surplus positions, and the nature of evolving trends.

6.2.1.5

With respect to the constraint observed above (section 6.2.1.4) it should be recorded that the DTI plans a benchmark inquiry into overseas payments and receipts for royalties, patents and licenses which is intended to provide data for 1987. The register of companies involved in such transactions will be enhanced by information obtained in the Census of Production inquiry. This expanded register will be used for the benchmark exercise. Results from the inquiry are expected to be available by early 1989, and it is hoped that this will provide the basis for grossed-up estimates of royalty payments and receipts by country.

6.2.2 *Royalties Transactions—Aggregate Values*

6.2.2.1

One table in each year's coverage of royalties (e.g. [QRL.10]; 1984 edition; table 5.1) provides estimates for all business (excluding oil, banking, insurance) i.e. including estimated coverage for non-respondents to the annual surveys. The tables provide data in respect of transactions with (1) related concerns overseas, (2) unrelated concerns overseas, (3) all overseas concerns (i.e. (1) plus (2)).

For each group of concerns information is provided on receipts, expenditure and the balance (i.e. receipts less expenditure). These totals for receipts, expenditure and balance are also subdivided into (1) artistic royalties and (2) technology and mineral royalties (see section 6.2.1.1).

6.2.2.2
A time series for the data described in section 6.2.2.1 can be compiled for 1964 to 1984. This can be achieved using the *Business Monitor* M4/MA4 [QRL.10] sources (see Appendix 8) for 1973 (table 37), for 1974 (table 37) and for 1984 (table 5.1).

6.2.3 *Royalties Transactions by Country*

6.2.3.1
In the 1984 coverage ([QRL.10]; 1984 edition; table 5.2) royalties data are provided broken down by area and country; though, as discussed in section 6.2.1.4, it needs to be noted that the figures given are only for 'returns received'. The table provides 1984 data for net royalty receipts derived from total royalty receipts less total royalty payments.

Total royalty receipts and total royalty payments are both subdivided in two ways. Firstly, into transactions between (1) 'related' concerns and (2) between 'unrelated' concerns. Secondly, into transactions relating to (1) 'printed matter, sound recordings or performing rights', (2) 'technological and mineral royalties'.

6.2.3.2
Data on the basis described in section 6.2.3.1 above are available in *Business Monitor M4/MA4* for all years from 1966 to 1984, and also for 1964 and 1965 except that the division by types of royalty is not available for those years. From 1965 to 1981 the data for receipts and expenditure are included in separate tables.

6.2.4 *Royalties Transactions by Industry*

6.2.4.1
In the 1984 coverage (e.g. [QRL.10]; 1984 edition; table 5.3) data are provided on receipts and expenditures (returns received only) broken down by industry. Here companies are classified by their main UK activity, even though this may not necessarily reflect the industrial activity to which the recorded transactions relate. The data for both receipts and expenditures are subdivided into those with 'related' concerns and 'unrelated' concerns.

6.2.4.2
Although data are available annually from 1964 to 1984, a substantial change in the industry coverage effectively divides this into two sub-periods. From 1972 to 1984 ten manufacturing industries are covered (eleven from 1979 on, when 'shipbuilding' receives separate coverage), plus 'distributive trades' and 'other activities'. Up to 1971 fewer industries were covered, with some variation in coverage in 1964 and 1965. Oil, banking and insurance are excluded throughout.

6.3 Film and Television Transactions

6.3.1 *Introduction*

6.3.1.1

The Department of Trade and Industry's annual inquiry in respect of film and television companies covers the value of transactions between UK and overseas residents relating to films of any width and recordings on tape or any other media. The total value of transactions covered may be subdivided into those relating to 'production' and those relating to 'performance'.

Performances in cinemas or on television include the outright sales or purchases of films, as well as receipts or payments for rentals.

Production of films includes overseas receipts or payments for film processing, printing, editing, and sums for work on location. Thus receipts for production in the UK comprised amounts remitted from overseas, or put up from amounts held in the UK on behalf of overseas residents, and sums applied to the repayment of loans raised in the UK. The comparable expenditure figure includes amounts remitted from the UK, or put up from sums held overseas on behalf of UK residents, to finance film production, including location work, abroad.

The recorded returns from film companies also include any overseas transactions they do on behalf of another producer, distributor or agent in the UK.

The profits of UK subsidiaries and branches in the film industry due to their overseas parent companies, and those due to the UK from the overseas subsidiaries and branches of UK film companies, are excluded, being included in the balance of payments under 'interest, profits and dividends'. Also omitted from these figures, and included under 'overseas direct investment', are investments not associated with film production by overseas parent companies in their related UK concerns in the film industry, and UK film companies' non-film investments in their related overseas concerns.

The data reported in the inquiry include some imports and exports of exposed cinema film, which are recorded in the overseas trade accounts. However, the trade account also includes some exposed film not covered by the inquiry. This includes film flown in from units on overseas location for processing and return to the unit abroad, and film imported and prints exported after processing by specialist laboratories, to the order of overseas customers. An approximate indicator of this overlap is that the inquiry results probably include about half of the export figure in the overseas trade account for such processed film, and a somewhat higher proportion of the import figure.

6.3.1.2

In the 1979 inquiry the reporting base for film transactions was extended to achieve a more comprehensive coverage of smaller companies. This change is discussed in *British Business* [QRL.7] for 19th December 1980 (pp. 794-5), whilst the basis of the register previously used is described in the *Board of Trade Journal* [QRL.5] for 6th January 1967 (p. 22). The extended sample in 1979 covered 308 film and television companies (plus the BBC), of which 164 had not been covered in the previous year's inquiry. The results of the 1979 survey indicated that previous inquiries had underestimated the overseas transactions of smaller companies. To make some allowance for this the film company data reported for 1979 incorporated revisions for earlier years. Further reference to these revisions are made in the discussion below as appropriate.

6.3.2 *Overseas Transactions in Film and Television Material*

6.3.2.1
Each of the sources (see Appendix 9) provides a summary table encompassing both film and television companies' overseas transactions. Total receipts (sums receivable from overseas residents) and total expenditure (sums payable to overseas residents) are both subdivided into those of (1) film companies and (2) television companies. Here film companies' transactions include any in respect of rights restricted to television showings. Television transactions include those in respect of BBC sound broadcasting and Independent Television News.

6.3.2.2
A time series of the data outlined in section 6.3.2.1 can be compiled covering the years 1963-1985. The optimum series, in terms of encompassing all the revisions (back to 1970) described in section 6.3.1.2, can be constructed from table 1 in *British Business* [QRL.7] 19th September 1986, *British Business* [QRL.7] 19th December 1980, *Trade and Industry* [QRL.60] 1st August 1974.

6.3.2.3
For film companies only table 1 in the *Board of Trade Journal* [QRL.5] for section 6th January 1967 permits the extension of the time series of 6.3.2.2 back to 1961.

6.3.3 *Cinema Films Analysed by Area and Type of Transaction*

6.3.3.1
Information in each source (e.g. table 6.3 in [QRL.10], 1984 edition; table 3 in [QRL.7] for 19th September 1986) divides film companies' receipts and expenditures by type and by geographical area. The two types of film companies transactions covered are those introduced in section 6.3.1.1, i.e.:
(1) Performances in cinema or on TV
(2) Production of films etc.

The areas currently delineated are (1) European Community, (2) Other Western Europe, (3) North America, (4) Other Developed Countries, (5) Rest of World. For this coverage the companies were asked to report amounts against the geographical area to which they were payable or from which they were receivable, irrespective of where they might have first been earned.

6.3.3.2

With some alterations in the areas covered a time series of such data can be compiled for 1964 to 1985. The revisions discussed in section 6.3.1.2 can be incorporated in this series from 1974 onwards.

6.3.3.3

Data are also provided (e.g. by table 6.3 in [QRL.10], 1984 edition; table 2 in [QRL.7], for 19th September 1986) on the receipts of film companies 'in respect of rights restricted to television'. No data are provided on corresponding payments for foreign films for television showing, but the amounts involved are believed to be small.

6.3.4 *Cinema Films Analysed by Type of Company*

6.3.4.1

In the 1984 *Business Monitor MA4* coverage ([QRL.10]; 1984 edition) table 6.4 divided film companies' receipts and expenditures by type of transaction and by type of company.

The two types of transactions covered are again those described in section 6.3.1.1, i.e.:

(1) Performances in cinema and on TV
(2) Production of films etc.

The two types of company covered are:

(1) UK subsidiaries of the major US film companies
(2) Other companies.

6.3.4.2

In recent years the *British Business* coverage (e.g. [QRL.7]; 19th September 1986; table 4) by type of company does not have the further division by type of transaction incorporated in the *Business Monitor MA4* coverage described in section 6.3.4.1.

6.3.4.3

From sources listed in Appendix 9 data on the basis described in section 6.3.4.1 can be compiled covering 1965 to 1984. The revisions discussed in section 6.3.1.2 can be incorporated in this series from 1974 to 1978.

6.3.5 *BBC and IBA Contractors Analysed by Area and Type of Transaction*

6.3.5.1
In the 1984 *Business Monitor MA4* coverage ([QRL.10]; 1984 edition) table 6.2 divides TV companies' receipts and expenditures by type of transaction and by geographic area.

Once again the types of transaction distinguished are those described in section 6.3.1.1 i.e.:
(1) Performances in cinemas or on television
(2) Production of films etc.

These transactions are further broken down by area, covering (1) European Community, (2) Other Western Europe, (3) North America, (4) Other Developed Countries, (5) Rest of World.

6.3.5.2
The degree of subdivision described in section 6.3.5.1 is only available in the *Business Monitor MA4* [QRL.10] coverage, and there only from the 1982 edition onwards (see table 6.2 in each edition). This provides a time series from 1975 to 1984 on this basis. However, allowing for some adjustments in the areas covered, some information on the geographical breakdown of the TV companies' transactions is available from sources listed in Appendix 9 from 1965 onwards.

6.4 Direct Exports

6.4.1 *Introduction*

6.4.1.1
The data reviewed in this section (unlike that surveyed in chapter 2) are derived predominantly from the annual overseas transactions survey carried out by the Department of Trade and Industry and its predecessors. Some of the tables reported also included data (on oil companies and other large enterprises not responding to the survey) derived from published sources, most notably company accounts. However, the provisions of the Companies Act 1981 meant that companies were no longer required to publish the value of their direct exports in their accounts. This obviously compromised that part of the coverage dependent upon such published sources. Beyond this, the removal of the obligation to publish such data meant that companies had less need to compile the information in a form suitable for response to the survey, thus further compromising the exercise. Therefore, the 1981 results are the last reported on direct exports, and even those are somewhat less comprehensive than in earlier years. Since some of the perspectives developed in the data remain of interest, we review below the data provided by this part of the overseas transactions inquiry, despite the lack of contemporary results. See Bosworth [B.5], Appendix 18, for the 1974 'overseas transactions' survey form incorporating the direct export questions.

6.4.1.2

The analysis concentrated on *direct* exports, defined as those exports for which the export document was submitted to Customs and Excise by either the enterprise making the return, its own exporting subsidiaries, or export agents working to its instructions. Such direct exports would not necessarily represent the whole of an enterprise's contribution to the export earnings of the UK since part of its output may be sold to other companies and in turn exported by them, with or without further processing.

6.4.1.3

The distinctive features of the overseas transactions inquiry information on exports (compared with aggregate trade figures) are the breakdown of such data by certain characteristics of the exporting firms. One of these firm characteristics is the size of exports (see section 6.4.2).

6.4.1.4

A second firm characteristic used in analysing direct exports concerns the nature of the UK firm's links with overseas enterprises. Thus five categories are distinguished:

(1) United States controlled UK enterprises
(2) Other foreign controlled UK enterprises
(3) UK associates of foreign enterprises
(4) UK enterprises with overseas affiliates
(5) UK enterprises with no overseas affiliates.

6.4.2 *Large Exporters Analysed by Size of Exports*

6.4.2.1

In the 1981 *Business Monitor MA4* coverage ([QRL.10]; 1981 edition) table 6.1 classifies enterprises into seven groups by 'size of direct exports during the year'. These groups range from 'under £1 million' to 'over £100 million'. The table provides annual data from 1971 to 1981. For each year data are given on the number of enterprises in each export size group, and the value of their exports. The figures include coverage of oil companies (see section 6.4.1.1), but excludes diamonds. The data also include estimates for non-respondents with exports of over £5 million (in the five largest size groups) from 1978 to 1981, for those with exports over £2 million from 1975 to 1977, for those over £1 million up to 1974.

6.4.2.2

Earlier sources (see Appendix 10) provide similar data for years 1970 and 1965 to 1967.

6.4.3 *Total Exports Analysed by Category of Enterprise*

6.4.3.1
In the 1981 *Business Monitor MA4* coverage ([QRL.10]; 1981 edition) table 6.2 divides exports between the five categories of enterprise introduced in section 6.4.1.4. Thus for each year, from 1971 to 1981, the percentage of total exports accounted for by each of these five categories of enterprise is given. Also (for 1973, 1976 and for 1979 to 1981) the percentage of total exports (i.e. aggregate of all exports of the five groups) accounted for by 'exports to related concerns' of each of the categories of firm is given. Though not given directly this facilitates the simple calculation of the orientation to 'intra-group-exports' of each of the types of enterprise. The data include oil companies, but exclude diamonds, and cover 'returns received adjusted approximately for non-response'.

6.4.3.2
Data for 1970, on the basis described in section 6.4.3.1, are available in table 3 in the *Trade and Industry* coverage for that year ([QRL.60]; 13th April 1972). This includes data on 'exports to related concerns'.

6.4.3.3
A quite familiar warning should perhaps be sounded at this point. This is that, for a number of often-discussed reasons, the value at which firms chose to record intra-group transactions (i.e. 'exports to related concerns' in the figures discussed here) may differ from the arms-length price that would prevail if the same goods were transferred between independent parties.

6.4.4 *Category of Enterprise and Main Industrial Activity*

6.4.4.1
In table 6.3 in the 1981 *Business Monitor MA4* coverage ([QRL.10]; 1981 edition) data for 1981 are broken down by category of enterprise (see section 6.4.1.4) and by industry group (six manufacturing industries, plus 'other activities'). Thus, for each category of enterprise, the table divides (1) total exports and (2) related exports, between industry groups.

The data exclude both oil and diamond companies, and only cover 'returns received' in the inquiry.

6.4.4.2
From the sources listed in Appendix 10 data similar to that discussed in section 6.4.4.1 (though with some variation in industry coverage) are available for 1966, and from 1970 to 1981. Of these years information on 'related exports' is given for 1966, 1970, 1973, 1976 and 1979 to 1981.

6.4.5 *Category of Enterprise and Size of Total Exports*

6.4.5.1

In the 1981 *Business Monitor MA4* coverage ([QRL.10]; 1981 edition; table 6.4) data for that year is broken down by category of enterprise (see section 6.4.1.4) and by 'size group of direct exports' (six groups are covered, from 'under £1 million' to 'over £50 million'). For each category of enterprise, (1) number of firms, (2) total exports, (3) related exports, are divided between the six size groups. The figures are limited to 'analysis of returns received', and exclude oil and diamond companies.

6.4.5.2

From the sources listed in Appendix 10 data similar to that discussed in section 6.4.5.1 (though with some variations in size categories) are available for 1966 and for 1970 to 1981. Of these years information on 'related exports' is given for 1966, 1970, 1973, 1976 and 1979 to 1981.

6.4.6 *Direct Exports Analysed by Terms of Original Credit*

6.4.6.1

From 1966 to 1978 the overseas transactions survey analysed direct exports by terms of any related credit. In 1978 terms of credit were divided under six classes, ranging from up to sixty days to over five years. Thus in table 6.3 in the 1978 *Business Monitor MA4* ([QRL.10]; 1978 edition) total exports are divided between these classes (for 'returns received adjusted approximately for non-response'). From sources listed in Appendix 10 similar data are available for each year from 1966 to 1978, although with some variations in categories of credit terms.

6.4.6.2

In the 1978 *Business Monitor MA4* coverage ([QRL.10]; 1978 edition) table 6.6 analyses exports in that year by terms of credit (six categories) and by industry (seven industries). Thus each industry's exports are broken down by terms of original credit. The table is limited to analysis of 'returns received', and excludes oil companies and public sector enterprises. With some changes in industry coverage and credit classes, similar data are available for each year from 1966 to 1978.

CHAPTER 7

INTERNATIONAL TRANSACTIONS OF THE UK MONETARY AND BANKING SECTOR

7.1 Introduction

The extensive and long-standing international involvement of many sectors of the UK economy has had the inevitable result of generating a similar international orientation in the UK financial sector. In fact, the financial sector must rank as one of the most internationally oriented sectors of the UK economy, and, similarly, the UK's financial sector must be one of the most externally oriented in the world. Systematic data on various aspects of the international transactions of the UK monetary and banking sector are to be found in the Statistical annexe to the *Bank of England Quarterly Bulletin* [QRL.4].

In this chapter we review those on-going series (as of the February 1987 edition: Vol.27 No.1) from this source, which include important information on the financial sector's international commitments. This is facilitated by the practice of retaining fairly consistent labelling of series in the *Bulletin*. Where significant changes in presentation have occurred within a series we note these. Similarly, major changes in the reporting base for series are noted here, although it should also be borne in mind that minor changes in coverage are noted consistently in the footnotes to the series. Detailed notes and definitions are provided once a year ([QRL.4]; February 1987 edition; March edition in earlier years), with changes in coverage noted in the relevant edition as and when they occur.

7.2 Balance Sheet Data

7.2.1 *Banks' Balance Sheets (Aggregates)*

A series of tables ([QRL.4]; tables 3.1 to 3.9 in February 1987 edition) provide data on the balance sheets of banks in the UK.

The first table (3.1) aggregates the balance sheets of the UK offices of those members of the monetary sector which report monthly. Generally, this covers those with a balance sheet of £100 million or more, or eligible liabilities of £10 million or more. Members of the London Discount Market Association (LDMA) are excluded, and covered separately in table 4.

(1) *Liabilities*. In the coverage of *sterling* liabilities the table ([QRL.4]; February 1987 edition; table 3.1) gives information on the 'overseas' share in (a) sight deposits, (b) time deposits (separately). In the coverage of *other currency* liabilities the table gives information on the 'overseas' share in sight plus time deposits (taken together).

(2) *Assets*. In the coverage of *sterling* assets the table gives information on the overseas share in (a) market loans, (b) advances. In the coverage of *other currency* assets the table gives information on the 'overseas' share in (a) market loans and advances, (b) investments.

Overseas market loans are defined to include all balances (including correspondent balances) with, and funds lent to, banks overseas (except trade and portfolio investment); bills drawn by banks overseas under acceptance credit facilities opened by the reporting institution which the reporting institution has itself discounted; and certificates of deposit, promissory notes, and other short-term paper issued by banks overseas and owned by the reporting institution.

Advances to overseas includes all lending for exports under ECGD bank guarantee (excluding any amounts refinanced).

Liabilities and assets are reported at the value standing in the reporting institutions' books, with those in currencies other than sterling being converted into sterling at the closing middle-market spot rate on reporting days. This means that some figures are therefore liable to be affected by exchange rate movements.

In all these tables *overseas residents* are defined to include overseas governments, together with their diplomatic and military offices and representatives in the UK; international organisations, including their branches or representatives in the UK; UK representative offices of overseas banks; overseas offices of the reporting institutions; enterprises located and operating overseas, including branches and subsidiaries of businesses which are themselves UK residents; and persons (including those of UK origin) who are residing outside the UK for a year or more. In each edition the table provides monthly data for the most recent year. In October 1986 the reporting day was changed from the 3rd Wednesday of each month to the month's last day.

The banking series were originally introduced in the May 1975 edition, but the introduction of the new monetary sector on 18th November 1981 resulted in a break in the series as shown in the tables of the March 1982 *Bulletin* [QRL.4]. Full definitions for the old series can be found in the December 1981 *Bulletin* (see also [QRL.4] December 1983, pp. 562-3).

The reporting population relevant to the series changes from time to time. The notes to the *Bulletin* [QRL.4] statistical annex regularly record the number of institutions joining and leaving the population, and also report the net effect of these changes on sterling and foreign currency assets. Similar data is reported in Table 6.4 of CSO *Financial Statistics* ([QRL.33]; February 1987 edition).

7.2.2 *Banks' Balance Sheets—Monthly Reporting Institutions*

7.2.2.1

Tables 3.2 to 3.8 provide balance sheet information on seven groups of monthly reporting institutions. These are:

Retail Banks (table 3.2) which comprises banks which either have extensive branch networks in the UK or participate directly in a UK clearing system, except for Standard Chartered Bank, which is included in Other British Banks, and

Citibank NA, which is in American Banks. It includes branches in the Channel Islands and the Isle of Man which the parent banks have opted to include within the UK monetary control arrangements and therefore within the UK monetary sector. The series based on this definition commenced in December 1983.

Accepting Houses (table 3.3) comprising the members of the Accepting Houses Committee and certain of their subsidiaries in the Channel Islands and the Isle of Man.

Other British Banks (table 3.4) which comprises all other UK-registered institutions and certain institutions in the Channel Islands and the Isle of Man which are either independent companies, or controlled by UK companies or by individuals. Banks which are controlled by overseas companies are classified as overseas banks, so that the subsidiaries of Allied Irish Banks Limited and the Bank of Ireland are included within the group of other overseas banks. The series based on this definition commenced in December 1983.

American Banks (table 3.5) comprising the branches and subsidiaries, including certain offices in the Channel Islands, of American banks. The banking branches and subsidiaries in the UK of American companies which are not banks in the US are included within the group of other overseas banks.

Japanese Banks (table 3.6) comprising the branches and subsidiaries of Japanese banks.

Other Overseas Banks (table 3.7) which comprises the branches and subsidiaries, including certain offices in the Channel Islands and the Isle of Man, of all foreign companies other than American and Japanese banks.

Consortium Banks (table 3.8) which comprise UK-registered institutions which are owned by banks or financial institutions but in which no one institution has a shareholding of more than fifty per cent, but in which at least one shareholder is based overseas.

The reform in the composition of these groups which occurred in September 1983 is explained in a note in the December 1983 *Bulletin* [QRL.4] (pp. 562-3).

7.2.2.2

Each of the tables 3.2 to 3.8 [QRL.4] presents balance sheet data to a consistent format. This includes:

Liabilities: in the coverage of liabilities the tables give information on the overseas share in (a) sterling deposits, (b) other currency deposits.

Assets: in the coverage of *sterling* assets the tables give information on the overseas share in (a) market loans, (b) advances. In the coverage of *other currency* assets the table gives information on the overseas share in market loans plus advances (taken together).

In tables 3.2 to 3.8 the data are reported monthly for the most recent year.

7.2.3 *Balance Sheets—Quarterly Reporting Institutions*

A new table ([QRL.4]; February 1987 edition; table 3.9), starting with data for the 3rd quarter of 1986, aggregates the balance sheets of the UK offices of those monetary sector institutions which report data on a quarterly basis only.

In the coverage of *liabilities* information is given on 'overseas' share in (a) sterling deposits, (b) other currency deposits. In terms of *sterling assets* data are given on reporting institutions 'balances with and loans to banks overseas', and also on the share of 'overseas' in 'advances'. In terms of *other currency assets* data are given on reporting institutions 'balances with and loans to overseas'.

7.2.4 *Discount Market: Balance Sheet*

Table 4 ([QRL.4]; February 1987 edition) shows the liabilities and assets of the discount market, which currently (February 1987 edition) comprises the eight discount houses that are members of the London Discount Market Association. Before the introduction of the new monetary sector on 18th November 1981, the market had also included the money trading departments of the five listed banks, but after this date the departments' figures have been included with those of their parents in Table 3.

The *liabilities* side of the table gives information on overseas share in (a) sterling liabilities, (b) other currency liabilities. In terms of *sterling assets* the table gives information on overseas share in funds lent. The data are reported for the most recent year. Similar data are given in table 6.5 in CSO *Financial Statistics* ([QRL.33]; February 1987 edition).

7.2.5 *UK Monetary Sector: Consolidated Balance Sheet*

The figures in table 6 ([QRL.4]; February 1987 edition) are designed to show the position of the monetary sector (including all those institutions covered by tables 3 and 4) as a whole in relation to third parties. Thus transactions between institutions within the sector are excluded (in a way that would not have been appropriate with the subsector data of tables 3.2 to 3.9 and table 4), and various adjustments are made to the figures on consolidation to avoid double-counting.

The *liabilities* side includes data on overseas sector deposits, divided into (a) sterling, (b) other currencies. The *assets* side includes data on lending to overseas sector, also divided into (a) sterling, (b) other currencies.

The introduction of the new monetary sector on 18th November 1981 created a break in this series, which is described in an article in December 1981 *Bulletin* ([QRL.4]; December 1981 p. 351). Similarly a major change in the reporting population occurred at end-March 1983, following the completion of the revision of banking statistics described in an article in March 1983 *Bulletin* ([QRL.4]; March 1983 pp. 69-72). The more regular on-going changes in population are reported as necessary in the manner described earlier (see section 7.2.1).

Data are reported quarterly for the most recent three years. Similar data are given in table 6.1 of CSO *Financial Statistics* ([QRL.33]; February 1987 edition).

7.3 Central Government Overseas Transactions

7.3.1 *Overseas Contribution to Government Borrowing Requirement*

7.3.1.1
The central government borrowing requirement (CGBR) is a measure of the net expenditure (including lending) of the central government resulting broadly from decisions of a budgetary nature. Table 7 of *Bank of England Quarterly Bulletin* ([QRL.4] February 1987 edition) analyses government borrowing, for recent quarters and a longer run of financial years, by the source of the funds borrowed. The financing of the CGBR is broken down into (1) external and foreign currency financing, and (2) domestic sterling finance. The former concerns us here.

7.3.1.2
The data in the table divide *external and foreign currency finance* of the CGBR into three components:
(1) Changes in official reserves (other than increases arising from new allocations of Special Drawing Rights)
(2) Net borrowing in foreign currency (including IMF drawings)
(3) Changes in other overseas holdings of central government debt. This is subdivided into (a) notes and coin, (b) marketable debt. Marketable debt, in turn, is divided into (i) stocks (ii) treasury bills.

Together these items represent the impact on the central government's domestic financing of the overseas transactions of the country as a whole.

7.3.2 *External Transactions in Marketable Government Debt*

7.3.2.1
The *Bank of England Quarterly Bulletin* data on transactions in marketable government debt ([QRL.4]; February 1987 edition, table 8) provides some further information on the overseas sector's contribution to these transactions.

7.3.2.2
The data on the net overseas sector purchases of *stocks* (category 3.b(i) in section 7.3.1.2 above) is divided between the purchases of (1) central monetary institutions and international institutions, and (2) other.

Here central monetary institutions (CMIs) cover banks, most currency boards, the Crown Agents when acting as agents of CMIs, the Bank for International Settlements, and a few other accounts known to be used for central monetary purposes.

It is noted that these figures for the overseas sector transactions in stocks are based partly on the statistics of UK external liabilities in sterling ([QRL.4]; February

1987 edition, Table 16; see 7.6 below) and partly on information extracted from the registers of government stocks; as such, it is warned, they should not be regarded as more than estimates.

7.3.2.3

The data in the *Bulletin* [QRL.4] table 8 divides the net overseas sector purchases of *treasury bills* (category 3.b(ii) in section 7.3.1.2 above) into those of (1) CMIs and international institutions, and (2) other. The change recorded is in the book value of the treasury bills.

7.4 External Liabilities and Claims of the UK Monetary Sector and Other Financial Institutions

7.4.1 *Introduction*

The *Bank of England Quarterly Bulletin* [QRL.4] data we have surveyed so far in this chapter provide evidence on the relative importance of overseas transactions; on the balance sheets of institutions (section 7.2) and on central government finance (section 7.3). The next five sections (sections 7.5—7.9) focus more directly, and in more detail, on such external activity.

7.4.2 *Institutions Covered*

In the *Bulletin* [QRL.4] table 14 surveys the external liabilities and claims of the UK monetary sector and other financial institutions. In addition to the usual members of the monetary sector the table also covers miscellaneous financial institutions. Currently ([QRL.4]; February 1987 edition) these include banks in the Channel Islands and Isle of Man which have opted out of the monetary sector, three non-monetary sector gold dealers and a small number of other financial institutions located on the mainland that are outside the monetary sector but make loans to, or take deposits from, overseas residents. The table now also includes data for smaller monetary sector institutions which complete less detailed returns.

Gross liabilities to, and claims on, overseas banks and other overseas customers are reported, as recorded in the reporting institution's books at the close of business on the last working day of each calendar quarter. Since the majority of these assets and liabilities are denominated in US dollars, the data in the table are reported in that currency rather than sterling. This reduces any distortion in comparisons that might be caused by fluctuations in exchange rates between reporting dates. Amounts of other currencies have been translated into dollars at the closing exchange rate for the reporting date.

The table now covers liabilities and claims denominated in sterling as well as those denominated in foreign currencies. An article introducing this change in coverage appeared in the *Bulletin* [QRL.4] for September 1983 (pp. 392-4). That article also provides retrospective annual coverage of the external liabilities and claims in sterling of the monetary sector and other financial institutions in the UK back to December 1979, with a detailed geographical coverage.

7.4.3 *UK Monetary Sector Liabilities*

The UK liabilities of the sector comprise deposits and advances received by reporting institutions from overseas residents, and certificates of deposit issued in London and held by overseas residents. Excluded from the series are official liabilities in foreign currencies which arise from direct borrowing by the UK government or from drawings on central bank facilities.

The liabilities section of the table comprises two parts. The first divides UK liabilities by currency. Coverage is provided for liabilities in (1) sterling, (2) US dollars, (3) other currencies; of which Deutschmarks, Swiss Francs, Japanese Yen are separately distinguished.

For each of these three currencies (treating other currencies in total) liabilities to (1) central monetary institutions, (2) other banks, (3) other non-residents, are separately distinguished. In addition, that part of the US dollar-denominated liabilities owed to the US is separately distinguished. Here data are provided quarterly for the five most recent quarters.

The second part provides a very detailed country breakdown of the UK liabilities. In this part, for each country, the foreign currency contribution to total (sterling plus foreign currency) liabilities is shown. Here data are provided quarterly for the three most recent quarters.

7.4.4 *UK Monetary Sector Claims*

The UK claims of the sector comprise the reporting institutions' loans and advances to overseas residents; commercial bills and other negotiable paper drawn on overseas residents; lending under ECGD special schemes for exports, including amounts refinanced; sterling acceptances given on behalf of overseas residents. With effect from end-December 1985, claims also include (i) assets leased out under finance leases, and (ii) holdings of certain overseas investments with an original maturity of one year or more. Those overseas investments include securities with contractually defined repayment dates issued by overseas governments, by overseas institutions such as the World Bank and by corporate bodies registered abroad, except subsidiaries of the reporting institutions. Excluded are investments in the share and loan capital of the reporting institution's own overseas subsidiaries, and in securities without contractually defined repayment dates, such as equities.

In order to allow for the incorporation of the two new categories of asset from December 1985, the tables show a break in the series at that date. Thus two sets of data are given for that month, one on the old basis and the other on the new.

The claims section of the table comprises two parts. The first part divides UK claims by currency. Coverage is provided for claims in (1) Sterling, (2) US dollars, (3) other currencies; of which Deutschmarks, Swiss Francs, Japanese Yen are separately distinguished. For each of the three currencies (treating 'other' currencies in total) claims on (1) Central Monetary Institutions, (2) other banks, (3) other non-residents, are separately distinguished. In addition, that part of the US dollar-denominated claims that are claims on the US is separately distinguished. Here data are provided quarterly for the five most recent quarters.

The second part of the table provides a very detailed country breakdown of the UK claims. In this part, for each country, the foreign currency contribution to total (sterling plus foreign currency) claims is shown. Here data are provided quarterly for the three most recent quarters.

7.5 Consolidated External Claims and Unused Commitments of UK-Registered Monetary Sector Institutions and their Worldwide Branches and Subsidiaries

7.5.1 *Introduction*

The reporting of these data ([QRL.4]; table 15, February 1987 edition) differs in coverage from that of other banking tables in two respects.

Firstly, the contributing organisations are limited to recognised banks and licensed deposit-taking institutions registered in the UK, comprising British-owned banks, consortium banks and those overseas banks registered in the UK. Thus the UK branches of banks incorporated abroad are omitted.

Secondly, the data do include the operations of overseas branches and subsidiaries of UK registered banks.

The data (given in dollars) cover one month (e.g end-June 1986 in the February 1987 *Bulletin*), and is presented with very detailed country coverage. Two components are distinguished (1) outstanding claims with a contractual repayment date, (2) unused external commitments.

7.5.2 *Outstanding Claims with a Contractual Repayment Date*

These comprise cross-border lending by UK offices in sterling and foreign currencies; cross-border lending by overseas branches and subsidiaries in local and non-local currency, other than to the UK; and lending by overseas branches and subsidiaries to local residents in non-local currency.

The data for end-June 1986 (reported in table 15 *Bulletin* February 1987) incorporate some additional claims. These include holdings of certain overseas investments with an original maturity of one year or more, comprising securities with a contractually defined repayment date issued by overseas governments, overseas institutions such as the World Bank and corporate bodies registered abroad (other than subsidiaries of the reporting institutions). Outstanding claims also include sterling and foreign currency acceptances given on behalf of overseas residents.

Another innovation from end-June 1986 is the inclusion of data on investments in securities without a contractually agreed repayment date (e.g equities) issued by overseas residents (other than subsidiaries of the reporting institutions). However, these figures are shown as a separate item and are not included in the total contractual debt.

In the table these claims are divided:

(1) by sector; i.e. claims on (a) banks, (b) public sector (including CMIs), (c) other

(2) by maturity; where maturities are classified by reference to the period remaining to maturity rather than to the original period of the loan.

7.5.3 *Unused External Commitments*
These are the unutilised portion of both binding contractual obligations and those commitments which reporting banks would regard themselves as obliged to honour in any circumstances.

They are divided by sector, into those of (1) banks, (2) public sector (including CMIs), (3) other.

7.6 External Liabilities and Claims in Sterling Reported by UK Monetary Sector and Other Financial Institutions

These sterling claims and liabilities (logically recorded in sterling, by comparison with the dollar-denoted data surveyed in sections 7.4 and 7.5) are reported in *Bulletin* table 16 ([QRL.4]; February 1987 edition). The data are provided on a quarterly basis. Earlier a break in these series had occurred in March 1983, due to a major change in reporting population (see section 7.2.5 above). The table is divided into five parts, these being described in sections 7.6.1 to 7.6.5.

7.6.1 *Central Monetary Institutions and International Organisations*
These cover readily marketable investments in the form of British government (or government-guaranteed) stocks, Treasury bills, and other banking and money-market assets held in the UK by central monetary institutions (including BIS) and international organisations.

The first part of this section of the table divides holdings into:
(1) Total banking and money-market assets, plus
(2) Non-interest bearing notes, plus
(3) British government stocks (these being revalued annually at current market prices).

Of these (1) is sub-divided into:
(i) deposits with banks etc.
(ii) deposits with local authorities
(iii) treasury bills

The second part breaks down total holdings geographically into those of (1) European Community, (2) Oil exporting countries, (3) Other developing countries, (4) Other countries, (5) International organisations (excluding IMF), (6) Unallocated.

The notes comment that "there are other holdings of sterling which countries may regard as part of their reserves; some are included in banking and money-market liabilities to other holders, and others are in types of asset-equities and eurosterling assets, for example, which are not covered by the table".

7.6.2 *Banking and Money-Market Liabilities to Other Holders*
These comprise deposits by overseas residents (including overseas offices of UK banks) with monetary sector and other financial institutions in the UK, and Treasury bills held by reporting institutions on behalf of non-residents, which are allocated to non-banks.

Thus the first part of this section of the table divides holdings into:

(1) deposits with UK banks placed by overseas banks
(2) deposits with banks placed by non-banks
(3) treasury bills.

The second part breaks down total holdings geographically into those of (1) European Community, (2) Other developed countries, (3) Eastern Europe, (4) Offshore banking centres, (5) Oil exporting countries, (6) Other developing countries, (7) Unallocated.

7.6.3 *Loans, Advances and Balances*

This section covers loans and advances to, and balances with, overseas residents by monetary sector and other financial institutions in the UK, and also acceptances by a reporting institution on behalf of overseas residents which have been discounted by that institution. From end-December 1985 coverage was extended to assets leased out under finance leases by these reporting institutions to overseas residents.

The total of loans, advances and balance is broken into claims on:

(1) central money institutions
(2) other banks
(3) non-banks

Geographically claims are broken down into those on (1) European Community, (2) Other developed countries, (3) Eastern Europe, (4) Offshore banking centres, (5) Oil exporting countries, (6) Other developing countries, (7) International organisations, (8) Unallocated.

7.6.4 *Bills and Acceptances*

These comprise commercial bills, promissory notes and other negotiable paper owned by the reporting institutions and drawn on overseas residents; acceptances given by the reporting institutions on behalf of overseas residents; and sterling lending under ECGD special schemes for exports, including amounts refinanced.

The data are divided into:

(1) ECGD—guaranteed paper
(2) other bills and acceptances.

Total claims are broken down geographically into those on (1) European Community, (2) Other developed countries, (3) Eastern Europe, (4) Offshore banking centres, (5) Oil exporting countries, (6) Other developing countries, (7) Unallocated.

7.6.5 *Securities*

The series on securities was added to the table, commencing with data for December 1985, as part of the revision in coverage which also affected loans, advances and balances (section 7.6.3). Securities comprise the reporting institutions' holdings of certain overseas investments with an original maturity of one year or more. These overseas investments include securities with contractually defined repayment dates issued by overseas governments, international institutions such as the World Bank, and by corporate bodies registered abroad. Investments in the share and loan capital of reporting institutions' overseas subsidiaries, and in securities, such as equities, with contractually defined repayment dates are excluded.

Data are broken down geographically into claims on (1) European Community, (2) Other developed countries, (3) Eastern Europe, (4) Offshore banking centres, (5) Oil exporting countries, (6) Other developing countries, (7) International organisations, (8) Unallocated.

7.7 Official Reserves

As reported by the Bank of England ([QRL.4]; February 1987; table 17.1) total reserves (in $) are divided into (1) Gold, (2) Special Drawing Rights, (3) Reserve position in the IMF, (4) Convertible currencies.

The table gives recent figures monthly (e.g since May 1985 in February 1987 edition), and longer time series (e.g since end-1978 in February 1987 edition) for end-year data.

7.8 UK Government and Other Public Sector Foreign Currency Debt

7.8.1 *Introduction*

These series (table 17.2 in February 1987 *Bulletin* [QRL.4]) are valued differently from the official reserves. Non-US-dollar components are translated into dollars (the table's reporting currency) at the closing market exchange rates on the last working day of each period to which they relate. The table distinguishes official debt and commercial debt.

7.8.2 *Total Official Foreign Currency Debt*

This category is divided into:

(1) IMF net drawings from the IMF, less other countries' net purchases of sterling, but excluding transactions which affect the UK reserve position in the IMF
(2) HMG short and medium term debt; divided into (a) securities (notes and bonds), (b) other
(3) HMG long-term debt
(4) Other public sector borrowing; divided into (a) that under exchange cover schemes, (b) uncovered borrowing.

7.8.3 *Total Public Sector Commercial Foreign Sector Borrowing*

This category is divided into:

(1) other public sector uncovered (commercial) borrowing
(2) leasing
(3) borrowing from overseas export credit agencies etc.

Recent data are given monthly (e.g. since April 1985 in February 1987 edition) and a longer series (e.g since end-year 1978 in February 1987 edition) for end-year data.

7.9 Balance of Payments 'Pink Book' Data

7.9.1 *Distinctive Facets of 'Pink Book' Data*

The annual CSO *UK Balance of Payments* 'Pink Book' [QRL.61], also provides data on the overseas transactions of the UK monetary and banking sector. The data are less detailed than that of the Bank of England source [QRL.4] but have the virtue of usually providing longer consistent time series; e.g annual data from 1975 to 1985 in most tables in the 1986 edition. Another major virtue of the 'Pink Book' coverage is that some of the tables put the transactions of the banking sector into a wider Balance of Payments context, in a way that the *Bank of England Quarterly Bulletin* [QRL.4] sources do not. Further the CSO 'Pink Book' provides information on the longer-term direct investment and portfolio investment transactions of the banking sector, which tend not to be covered by the Bank of England sources.

7.9.2 *Banking Sector*

The contribution of various banking sector transactions to the UK capital account are included in Table 1.4 in CSO *UK Balance of Payments* ([QRL.61]; 1986 edition). Table 8.1 also provides a capital account context for some banking sector transactions. Another 'Pink Book' table that places banking sector transactions in a wider context is that ([QRL.61]; Table 6.1; 1986 edition) on "overseas earnings of UK financial institutions". Here the earnings contribution of UK banking (monetary sector and miscellaneous financial institutions) is surveyed alongside other "City" institutions (e.g insurance: commodity trading; brokerage etc.).

The banking sector credit items covered in table 6.1 are:

(1) Financial services (see also table 3.6)
(2) Direct investment income due from overseas subsidiaries, branches etc. (see also table 5.2A)
(3) Portfolio investment income (see also table 5.3)
(4) Interest and discount on export credit
(5) Interest and discount on other lending in foreign currencies (see also table 5.1)
(6) Interest and discount on other lending in sterling (see also table 5.1)

The UK banking sector debit items covered are:

(1) Direct investment income due to, and services rendered by, overseas affiliates (see also tables 3.6 and 5.2A)
(2) Interest on borrowing in foreign currencies (see also table 5.1)
(3) Interest on borrowing in sterling (see also table 5.1)

7.9.3 *Assets and Liabilities of Banking Sector—Annual Transactions*

Section 8 of the CSO *UK Balance of Payments* ([QRL.61]; 1986 edition) covers annual changes in various assets and liabilities in the capital account. The banking sector's contribution to inward and outward direct investment is given in table 8.2A; whilst the portfolio investment overseas of the sector (since 1981) is in table 8.3. Table 8.4 concentrates on UK bank transactions, other than direct and portfolio investment.

Transactions in *external assets* of UK comprise:

(1) Foreign currency lending abroad by UK banks; divided into:

(a) identified long-term export credit
(b) other foreign currency lending

(2) Sterling lending abroad by UK banks; divided into:
(a) identified long-term export credit
(b) other identified export credit
(c) other sterling lending

Transactions in *UK liabilities* to overseas residents comprise:

(3) Foreign currency borrowing and deposit liabilities abroad of UK banks
(4) Sterling borrowing and deposit liabilities abroad of UK banks; divided into:
(a) overseas authorities' exchange reserves in sterling
(b) other liabilities

Net transactions are then reported divided into (a) foreign currency, and (b) sterling. An important discontinuity in reporting is involved concerning foreign currency transactions (2b and 3). Up to and including 1978 these transactions are reported net, under liabilities (i.e. 3). From 1979 onwards they are reported gross, under 2b and 3.

7.9.4 *Assets and Liabilities of Banking Sector—Annual Balance Sheet Data*

Section 9 of the *UK Balance of Payments* ([QRL.61]; 1986 edition) complements section 8 by giving end-year balance sheet (i.e. stock) data on assets and liabilities. Thus the banking sector's share of direct investment assets and liabilities is given in table 9.2A, and the share of portfolio assets (since 1981) in table 9.3.

Table 9.4 surveys the asset and liabilities position of the UK bank sector at each end-year, excluding direct and portfolio investment. The structure of the coverage is precisely that for annual transactions in table 8.4, described in section 7.9.3 above. However, the important discontinuity noted for the transactions data in section 7.9.3 does not affect these balance sheet figures.

CHAPTER 8

STERLING EXCHANGE RATES

8.1 Introduction

In this chapter we review sources of information on the exchange rate of sterling. The current rates are, of course, well documented in the financial press. Here we concentrate on those sources providing reasonably long time series of data. Although the market determined rate of exchange between sterling and another currency at a point in time is an unequivocal concept, various ways of reporting this for longer periods exist (e.g. end of period, period average etc.). For this reason we review in some detail below the format of presentation in each of the main sources. It should also be noted that the reported rates are the average of buying and selling rates, although these move in response to market forces.

Most sources providing information on the sterling exchange rate now include data on the sterling exchange rate index (ERI). This measure attempts to provide a summary of the movement of sterling against the range of other major currencies, rather than simply against the dollar, as had proved adequate prior to the general floating of currencies from 1971.

The effective sterling ERI, calculated by the Bank of England, is based on the weighting system of currencies derived in the IMF's Multilateral Exchange Rate Model (MERM). The MERM weights are such that any combination of changes in other currencies against sterling which would have the same effect on the UK trade balance as a one per cent change in sterling against each of the other currencies is reflected as a one per cent change in the index ([QRL.4]; March 1981 edition, p. 69). The MERM weights take account of the size of trade flows as well as of the relevant price elasticities and the feedback effects of exchange rate changes on domestic costs and prices, see Rhomberg [B.22] and Artus and McGuirk [B.2].

The sterling ERI originally introduced was described in detail in an article in *Economic Trends* [QRL.30] in June 1974. Revisions implemented in March 1977 are described in the issue of *Economic Progress Report* [QRL.28] for that month. Further revisions introduced in February 1981 are described in *Economic Progress Report* [QRL.28] for that month, and in the March 1981 edition of *Bank of England Quarterly Bulletin* [QRL.4]. Finally, improvements in the publication of sterling ERI are outlined in *Economic Progress Report* [QRL.28] for October 1984.

8.2 Sources

8.2.1 *CSO Financial Statistics Data*

The CSO publication *Financial Statistics* [QRL.33] provides extensive exchange rate information in table 13.1 (June 1987 edition) as follows:

(a) The sterling ERI (see section 8.1.2) is given, based on 1975 = 100.

(b) Information is provided on a sterling exchange rate index against European Monetary System (EMS) currencies. This is calculated in a manner similar to that of the broader ERI, but uses exchange rates for only the eight currencies which participate in the EMS exchange rate mechanism, rather than the seventeen foreign currencies included in the full index. Again, 1975 = 100 for this index.

(c) The table records average daily exchange rates (units of foreign currency per pound) of thirteen major currencies against sterling.

Information is given for the indices and rates in terms of the annual average for each of the last five complete years, and the monthly averages for the last four years.

In table 13.2 in *Financial Statistics* [QRL.33] information is provided on the sterling exchange rate with fifteen major currencies. For each foreign currency three pieces of information are given:

(a) The closing spot rate (units of foreign currency per pound) on the last working day of each month. Spot deals are those due to be executed for settlement within two working days.

(b) The three month forward margin for the foreign currency against sterling, expressed in units of the foreign currency.

(c) The three month forward margin for the foreign currency against sterling, expressed as a percentage per annum.

The data are given monthly for the most recent two years.

Finally table 13.3 [QRL.33] records sterling exchange rates against Special Drawing Rights (pounds per SDR) and against European Currency Units (pounds per ECU). End of year data is given for the last five years, and end of month data for the last four years.

8.2.2 *Bank of England Quarterly Bulletin Data*

Exchange rate information is provided by the *Bank of England Quarterly Bulletin* [QRL.4] in table 18 (March 1987 edition). The first part of the table gives effective exchange rate indices (see section 8.1.2) for sterling and eight other leading currencies. The second part presents information on exchange rates against sterling and the dollar of seven other leading currencies. The final part of the table includes data on US dollars in London. This covers the spot exchange rate (dollars per pound), and the three month forward premium expressed in cents and as an annual percentage.

The information in the first two parts of this table is given for:

(a) annual average for the last five completed calendar years

(b) last working day data for each of the last five years

(c) last working day data for the last four months

(d) Friday data for each week of the last four months. In the third part of the table only (b) to (d) are available.

8.2.3 *Economic Trends Data*
In *Economic Trends* ([QRL.30]; June 1987 edition) table 50 gives sterling exchange rates (average of daily rates in London) against six leading currencies (units of foreign currency per pound). The sterling ERI is also given (1975 = 100). Data are given for the most recent nine complete calendar years, quarterly for the last four years, and monthly for three years.

8.2.4 *CSO UK Balance of Payments Data*
The *United Kingdom Balance of Payments* ([QRL.61]; 1986 edition) provides limited exchange rate information, but for a longer period than other sources, i.e. annually since 1970. The information given is:
(a) US dollars per pound (end of period).
(b) US dollars per pound (annual average (i.e. average of trading day rates)).
(c) SDRs per pound (end of period).
(d) Sterling exchange rate index (annual average (1975 = 100)).

8.2.5 *IMF International Financial Statistics Data*
The UK country entry in the IMF publication *International Financial Statistics* [QRL.37] provides time series of seven measures of the exchange rate of sterling:
(a) SDRs per pound (market rate for end of period (i.e. year, quarter, month)).
(b) US dollars per pound (market rate for end of period).
(c) US dollars per pound (period average market rate).
(d) European Currency Units per pound (end of period rate).
(e) European Currency Units per pound (period average).
(f) Index of exchange rate, i.e. US dollars per pound (c above) expressed as an index with 1980 = 100
(g) Effective Exchange Rate. Calculation of the effective exchange rate of sterling (see section 8.1.2) expressed as an index with 1980 = 100.

The UK entry in *International Financial Statistics* gives annual data on these exchange rates for the most recent seven years, quarterly data for thirteen quarters, and monthly for ten months ([QRL.37] June 1987 edition).

In addition summary exchange rate tables in *International Financial Statistics* provide three further measures for sterling:
(h) Pounds per US dollar (daily market rate figures for the most recent available month i.e. two months prior to the date of edition).
(i) Pounds per SDR (daily rates for the most recent available month).
(j) Pounds per SDR (end of period market rate for fifteen most recent months).

CHAPTER 9

UK FOREIGN AID

9.1 Introduction

The UK Overseas Development Administration compiles and publishes detailed information on the UK performance in provision of Foreign Aid and related Technical Assistance. The information available from the Overseas Development Administration (including coverage of those parts of the aid effort which it does not administer) is reviewed in sections 9.2 to 9.8 of this chapter. International sources, comparing the UK aid effort to that of other countries, are surveyed in section 9.9.

9.2 UK Foreign Aid—Main Components

9.2.1 *Introduction*

The staple source for detailed information on UK aid provisions is the annual (the 1986 edition was the 21st) Overseas Development Administration publication *British Aid Statistics* [QRL.6]. The 1986 edition, scrutinised here, was the 4th to conform to the current, broadly consistent, format, although much of the information can be built into longer time series from earlier editions. The main data in the volume are presented in current prices, although indicators of constant price values are given in the introduction.

The Overseas Development Administration's annual descriptive review *British Overseas Aid* [QRL.8] also incorporates some of the summary information reviewed below in its statistical appendix.

9.2.2 *Net Aid Programme*

The major broad aggregates of aid provision (for 1981-85), and the relationships between them are set out in table 3 of the 1986 edition of *British Aid Statistics* [QRL.6]. The first of these aggregates is the *net aid programme* (gross expenditure less principal repayments), this being the central government budget for overseas aid administered by the Overseas Development Administration. It is divided into a bilateral aid programme (see section 9.3.1.2) and a multilateral aid programme (see section 9.3.1.6).

9.2.3 *Net Public Expenditure on Overseas Aid*

The next aggregate derived in table 3 is *net public expenditure on overseas aid* (PE). This adds five components to the net aid programme. These are:

(a) Commonwealth Development Corporation (CDC) overseas investment
(b) Debt rescheduling
(c) Extra costs of Ethiopian airlift (1984/85 only)
(d) Administrative costs of aid
(e) Other official development assistance multilateral contributions.

On the other hand, four components are deducted from the net aid programme to reach the net public expenditure on overseas aid, i.e.:

(f) Exchequer advances to CDC, i.e. annual flows from the aid programme to the CDC
(g) Repayments to the Consolidated Fund. This covers repayments of exceptional loans made to developing countries from the government's central consolidated fund rather than from the aid programme.
(h) Aid to developed countries; mainly the technical co-operation programme to black South Africans, and some disaster relief in developed countries
(i) Non-official development assistance multilateral contributions.

The table provides data for each of these nine components. Thus net public expenditure on overseas aid covers all UK government expenditure on overseas development; derived by adding certain aid oriented expenditures not administered as part of the overseas development administration's aid programme and by deducting certain aid programme expenditures, which either would result in double-counting (i.e. item (f)) or do not meet normal aid criteria.

9.2.4 *Total Official Aid Flows*

These coverage of public expenditure (PE) on overseas aid differs from the international concept of *total official flows* (TOF), (a) by excluding expenditure by the UK Export Credit Guarantee Department (ECGD), and (b) in the manner of timing the recording of some contributions to multilateral lending agencies.

(a) The ECGD is an official body, whose expenditures are aimed to assist and facilitate international commercial trade, predominantly UK exports. That part of its expenditure which is considered to benefit developing countries in this way enters the *other official flows* (OOF) component of total official flows, but is excluded from PE as not meeting the broader concept of developmental expenditures.

In Table 3 ([QRL.6]; 1986 edition) data are provided on two types of ECGD transactions, which facilitate the derivation of TOF from PE:

(i) Interest stabilisation grants. These are payments by the ECGD to commercial banks to enable them to maintain internationally-agreed interest rates for export credits for trade with developing countries.

(ii) Official export credits. This relates to a scheme which ran from 1972 to March 1980, under which the ECGD made loans to commercial banks to refinance an agreed portion of their longer-term credits to developing countries. Since the termination of new lending under the scheme the reported figures cover annual repayments.

(b) The timing difference between PE and TOF relates to UK contributions to the World Bank, the Regional Development Banks and the International Fund for Agricultural Development. These are made in the form of promissory notes, which are recorded in PE when the notes are cashed by the agency (encashments) and in Official Development Assistance (ODA) when the notes are deposited with the Bank of England (deposits). Table 3 ([QRL.6]; 1986 edition) provides annual data on 'deposits of promissory notes' and 'encashments of promissory notes', the adjustment from PE to TOF presentation being achieved by adding deposits to PE and subtracting 'encashments'.

9.2.5 *Official Development Assistance and Other Official Flows*
Finally Total Net Official Flows (TOF) is separated into Official Development Assistance (ODA) plus Other Official Flows (OOF). *Official Development Assistance* comprises flows to developing countries that are:
(i) from official funds
(ii) intended to promote development
(iii) on concessional terms (see section 9.3.4.2).
Other Official Flows cover official flows of development-oriented loans that do not qualify as ODA, including grants by ECGD for interest stabilisation, official export credits and CDC equity investment.

9.3 UK Foreign Aid—Gross Expenditure

9.3.1 *Bilateral and Multilateral Aid Flows*

9.3.1.1
A summary of *Gross UK Public Expenditure on Overseas Aid* is provided in table 4 of *British Aid Statistics* ([QRL.6]; 1986 edition). The first sub-division of data in this table is into (1) Bilateral and (2) Multilateral Flows.

9.3.1.2
Basically, *Bilateral Aid Flows* are those which are provided directly by the donor to a developing country, including expenditure in the UK such as that on students and trainees. Bilateral Aid is itself sub-divided in the table into three components (a) Financial Aid, (b) Technical Co-operation (TC), (c) Administrative Costs.

9.3.1.3
The data on Financial Aid is divided into six components, the first five of which come from the budget of the Overseas Development Administration:
(a) Project Aid (excluding Aid and Trade Provision)
(b) Aid and Trade Provision
(c) Programme Aid
(d) Debt Relief

(e) Other Non-Project Aid
(f) CDC Project Aid.

9.3.1.4

Technical co-operation (TC) is the provision of know-how, financed by grants. This may take the form either of the provision of advice or expertise to developing countries through the services of consultants, advisers or technical officers, or by the training of nationals from developing countries either in their own country or abroad. Data are given on six types of TC which are allocatable by country. These are:
(a) wholly-financed personnel
(b) partly-financed personnel
(c) students and trainees
(d) consultancies
(e) equipment and supplies
(f) pensions.
In addition, under TC data are given on four 'other programmes', namely:
(g) British Council
(h) research and development
(i) surveys
(j) other

9.3.1.5

The administrative costs, included in Bilateral Aid, cover the costs of running the Overseas Development Administration headquarters and Development Divisions of Foreign and Commonwealth Office staff engaged on aid work, and part of CDC administration. These costs are included with bilateral aid because the majority of them relate to the bilateral programme.

9.3.1.6

Multilateral Flows are those channelled through international bodies which promote social and economic development. In the case of those international bodies whose activities benefit both developing and developed countries, only that part of the UK contribution that is estimated to be development-oriented is counted as aid.

In *British Aid Statistics* Table 4 ([QRL.6]; 1986 edition) divides UK contributions to multilateral organisations into those to:
(i) European Community
(ii) World Bank Group
(iii) UN Agencies
(iv) Regional Development Banks
(v) Other

9.3.2 *Project Aid and Non-Project Aid; Terms of Aid*

9.3.2.1
A further sub-division of Gross UK Public Expenditure on Overseas Aid, by *purpose*, is incorporated in table 7 in *British Aid Statistics* ([QRL.6]; 1986 edition). The new information here sub-divides, by purpose, the Financial Aid component of Bilateral Aid (see section 9.3.1.2). The basic sub-division of bilateral financial aid is into project aid and non-project aid.

9.3.2.2
Project Aid is provided to establish new, or expand existing, production and infrastructure facilities. Data are provided on three sources of project aid within UK bilateral financial aid:
(i) from the Overseas Development Administration, through its regular country programmes
(ii) from the Aid and Trade Provision (ATP) of the Overseas Development Administration. The ATP is an allocation of funds used principally to enable British firms to match offers of mixed credit made by other countries. The ATP financial package consists of the aid element, usually a grant, associated with ECGD-backed export credits.
(iii) from the CDC, through its loans and investments.

9.3.2.3
Data are provided in the table on eight categories of *non-project aid*. These are:
(a) programme aid, which provides support for import financing to assist countries with particularly acute balance of payments problems
(b) budgetary aid, provided to help reduce the budgetary deficit of recipient countries. Information on allocation of budgetary aid by country is given in table 21 ([QRL.6]; 1986 edition)
(c) debt cancellation (see section 9.5)
(d) debt refinancing (see section 9.5)
(e) debt rescheduling (see section 9.5)
(f) food aid, provision of food supplies in kind. Information on allocation of food aid by country is given in table 22 ([QRL.6]; 1986 edition)
(g) disaster relief, emergency assistance other than food aid. Information on allocation of disaster relief by country is given in table 23 ([QRL.6]; 1986 edition)
(h) fertiliser aid, provision of fertilisers under the auspices of the FAO International Fertiliser Supply Scheme and the FAO Fertiliser Programme. Information on allocation of fertiliser aid by country is given in table 24 ([QRL.6]; 1986 edition).

The most recent year's information (e.g. 1985 in issue dated summer 1986) on food aid, budgetary aid and disaster relief, by recipient country, can also be found in table 5 in the statistical appendix to the Overseas Development Administration publication *British Overseas Aid* [QRL.8].

9.3.2.4

A second part to Table 7 in *British Aid Statistics* ([QRL.6]; 1986 edition) sub-divides Gross UK Public Expenditure on Overseas Aid, by *terms*. This division, by terms, relates only to the bilateral part of aid expenditure, and is itself divided into two main sub-sections: (a) grants and (b) loans and similar flows.

Grants are sub-divided as: (i) financial, (ii) technical co-operation, (iii) administrative costs. Four categories of *loans and similar flows* are distinguished: (i) interest-free loans, (ii) inter-governmental interest-bearing loans, (iii) CDC interest-bearing loans, (iv) CDC equity participation.

9.3.3 *Geographical Distribution of Bilateral Aid*

Summary information on the geographical distribution of Gross UK Bilateral Aid is presented in *British Aid Statistics* ([QRL.6]; 1986 edition; Table 5).

The first categorisation in the table divides recipient countries by income group. The groups delineated are:

(a) poorest fifty countries (excluding dependencies and countries with a population of less than 100,000)
(b) other low income countries; encompassing countries with GNP per capita of \$800 or less in 1983
(c) lower middle income countries; GNP per capita from \$800 to \$1400 in 1983
(d) upper middle income countries; GNP per capita of \$1401 or more in 1983

As an alternative figures are also provided on aid to the least developed countries (LLDCs).

The second geographical categorisation divides gross bilateral aid between (i) Commonwealth members (ii) non-Commonwealth. Data are also given on Dependencies, as a sub-group of Commonwealth members. The third geographical categorisation provided is by Continent. Finally, data are provided on leading individual aid recipients. The seventeen countries covered in the 1986 edition, are those which had occupied a position in the top ten recipients in one of the years 1981 to 1985.

9.3.4 *Aid Commitments—Grant Elements of Commitments*

9.3.4.1

Aid flows can be recorded at the time they are agreed, or at the time they are actually spent. The data in the tables so far reviewed are on the *expenditure* basis, i.e. the amount of money debited to the UK Exchequer during the reporting year. Alternatively, *commitments* are the sums recorded in written agreements or letters exchanged between donor and recipient country during the reporting year, specifying the terms and conditions on which aid is to be given.

9.3.4.2

In *British Aid Statistics* Table 6 ([QRL.6]; 1986 edition) presents data on the commitments basis. The major part of this table presents data in a format identical to that described in table 7 (see section 9.3.2). An important addition, however, is data on the average grant element of commitments. The concessionality of a loan depends on (1) its *interest rate*, (2) the *grace period* between the commencement of the loan and the date on which the first repayment of principal is due, and (3) the *number of years* over which repayments of principal must be made. The composite effect of these three elements can be measured by calculating the discounted present value of the future annual flow of principal and interest payments, and comparing this present value with the face value of the loan. The *grant equivalent* of the loan is then the difference between the discounted present value and the face value, whilst the *grant element* is the ratio of the grant equivalent to the face value. For a loan to qualify as ODA the grant element needs to be at least 25 per cent, using a 10 per cent discount rate.

The table provides data on the average grant element of:

(1) CDC loans and investments
(2) total loans
(3) total bilateral aid
(4) total aid.

9.3.5 *Contributions to Multilateral Agencies*

9.3.5.1

Extensive detail on the UK contribution to multilateral agencies is provided in table 8 of *British Aid Statistics* ([QRL.6]; 1986 edition). Six groups of agencies are covered:

(1) UN agencies; these being divided into (a) contributions to UN development agencies, of which twelve are distinguished and (b) contributions to the technical collaboration activities of other UN agencies, of which nine are distinguished.
(2) World Bank Group; of which the International Development Association is the most significant.
(3) Regional Development Banks; with ten Banks or Funds being distinguished.
(4) European Community; with nine Funds, Relief Efforts etc., being distinguished.
(5) Research Organisations; with four agricultural institutes etc., plus medical research, being distinguished.
(6) Other international organisations.

9.3.5.2

The most recent year's information (e.g. 1985 in the issue dated summer 1986) on the basis described in section 9.3.5.1 above can be found in the statistical appendix to the Overseas Development Administration publication *British Overseas Aid* [QRL.8].

9.3.6 *Project Aid by Sector*

9.3.6.1
Details of UK bilateral project aid is provided in two tables in *British Aid Statistics* ([QRL.6]; 1986 edition). Table 9 provides data on the commitments basis (see section 9.3.4.1), and table 10 on the expenditure basis. Though many projects may affect more than one sector of the recipient economy, it is usually possible to allocate a particular project by its main beneficiary sector. The tables, therefore, divide allocatable UK project aid between nine chief sectors, with several of these being quite extensively sub-divided.

The nine chief sectors delineated are:
(1) Renewable natural resources; with five sub-sectors
(2) Mining; with two sub-sectors
(3) Manufacturing; with three sub-sectors
(4) Energy
(5) Construction
(6) Trade, hotels and tourism
(7) Transport and communications; with seven sub-sectors
(8) Financial and miscellaneous services; with two sub-sectors
(9) Social and community services; with six sub-sectors

A certain number of projects were not allocatable to economic sectors and the tables provide data on these. The two main categories of non-allocatable projects were:
(1) projects in dependencies and associated states
(2) projects in recently independent and Caribbean and Pacific countries.

9.3.6.2
Data on project aid expenditure are given for the most recent year (1985 in issue dated summer 1986), arranged according to the format outlined in section 9.3.6.1, in statistical appendix table 4 in *British Overseas Aid* [QRL.8].

9.4 UK Foreign Aid—Technical Co-operation Expenditure

9.4.1 *Types of Technical Co-operation*
An important component of aid programmes takes the form of expenditures on technical co-operation (TC). Extensive detail of UK efforts of this type is provided in *British Aid Statistics* ([QRL.6]; 1986 edition; Table 11). Thirteen types of such expenditure are delineated in the table:
(1) *Personnel Overseas* expenditures (salaries, passages, allowances, training etc.) in support of personnel involved in technical co-operation activities overseas. Three main sub-divisions are delineated:
(a) wholly-financed personnel—personnel wholly-financed from the UK aid programme. The four components of this for which data are provided are: (i) long-term TC officers (ii) short-term advisers (iii) educational aid schemes (iv) training for British experts;

(b) partly-financed personnel; various supplementary allowances to individuals from Britain in the employment of overseas governments;
(c) British Volunteer programme; grants to bodies sponsoring volunteers.

(2) *Education and Training* two sub-divisions are delineated
(a) students and trainees; the major component of this comprises direct expenditure (cost of passages, scholarships, etc.) on students and trainees in Britain, and grants for training in country of origin or a regional centre. In addition, data are given on expenditure on World University Service, Commonwealth Education Co-operation, Fee support scheme;
(b) other education and training; the major component of this comprises grants to the British Council for administering the training programme, curriculum development, provision of libraries etc. Other functions under this heading for which data are given include, the cost of producing low-priced text books for use in developing country schools; schemes to present books to libraries and universities in developing countries; grants to universities or colleges in Britain running special courses for overseas students.

(3) *Research and Training* grants to institutions in Britain to undertake research about, and provide training for, developing countries. Data are provided relating to (i) Institute of Development Studies (ii) Centre for Tropical Veterinary Medicine (iii) other grants for research and training.

(4) *Research and Development* six areas of research and development for the benefit of developing countries are delineated:
(a) agricultural, forestry and fishing research
(b) energy research
(c) engineering research; with four sub-divisions
(d) medical research
(e) economic and social research; with three sub-divisions
(f) other research

(5) *Surveys* data are given on expenditure in support of (i) Overseas Surveys Directorate (ii) British Geological Survey (iii) Land Resources Development Centre (iv) Other Surveys.

(6) *Consultancies* payments for services provided by consultants other than those included under financial aid as part of the costs of a specific project.

(7) *Aid and Trade Provision Technical Co-operation* data provided separately from that for 'consultancies' from 1983 onwards only.

(8) *Equipment and Supplies* minor items of equipment to assist people working overseas under the aid programme.

(9) *Pensions and Compensation* such payments to individuals from Britain formerly employed on a supplemented basis.

(10) *Grants to voluntary organisations* in support of their assistance to developing countries.

(11) *Finance to other UK bodies assisting developing countries.*

(12) *Grants to non-governmental international bodies* the two main organisations on which data are provided are the International Committee of the Red Cross, and the International Planned Parenthood Federation.

(13) *Miscellaneous.*

9.4.2 *Alternative Data Source*
The most recent year's information (1985 in issue dated summer 1986) on the basis described in section 9.4.1 can also be found in table 3 in the statistical appendix to *British Overseas Aid* [QRL.8].

9.5 UK Foreign Aid—Debt Relief

An important element of the development assistance effort in recent years has taken the form of debt relief, i.e. attempts to lighten the burden of developing countries which find difficulty in meeting the terms of past loans. Three types of debt relief are distinguished:

(1) *Debt Cancellations* cancelling all outstanding payments of principal and interest
(2) *Debt Rescheduling* delaying the date on which principal and/or interest payments are due
(3) *Debt Refinancing* extending a new loan or grant to enable the debtor country to meet the debt service payments on earlier loans

In *British Aid Statistics*, table 20 ([QRL.6]; 1986 edition) provides data on these types of debt relief, by country. Where appropriate, total debt relief is sub-divided between that relating to principal repayments and to interest payments.

9.6 Commonwealth Development Corporation

9.6.1 *Introduction*
Information on the transactions of the Commonwealth Development Corporation, by country, are given in table 25 in *British Aid Statistics* ([QRL.6]; 1986 edition). Data are given for annual expenditure: (i) investment, (ii) capital repayments and realisations and (iii) investment income.

9.6.2 *CDC—Project Level Information*
Information on the activity of the CDC at the project level is given in *Commonwealth Development Corporation—Report and Accounts* [QRL.22].

9.7 UK Foreign Aid—Gross Bilateral Expenditure by Recipient Country

9.7.1 *Components of Bilateral Aid given by Country*
Detailed information on UK Gross Bilateral Aid Expenditure, by recipient country, is provided in table 18 in *British Aid Statistics* ([QRL.6]; 1986 edition).

The main aggregates on which data are given are:

(1) Total aid, including technical co-operation
(2) Official development assistance
(3) Total financial aid
(4) Total technical co-operation

It should be noted that total aid and ODA include administrative costs in addition to the total of financial aid and technical co-operation.

Five components of ODA financial aid are covered:
(1) Project aid (excluding aid and trade provision)
(2) Aid and trade provision
(3) Programme aid
(4) Debt relief
(5) Other non-project

Information is also given on CDC project aid.

Technical co-operation expenditure is divided into six components:
(1) Wholly financed personnel overseas
(2) Partly financed personnel overseas
(3) Students and trainees
(4) Consultancies
(5) Equipment and supplies
(6) Pensions.

9.7.2 *Alternative Data Source*

The most recent year's information (e.g. 1985 in issue dated summer 1986) on the basis described in section 9.7.1 can also be found in table 1 in the statistical appendix to *British Overseas Aid* [QRL.8].

9.8 UK Foreign Aid—Technical Co-operation, Manpower Data

9.8.1 *Introduction*

A series of tables in *British Aid Statistics* ([QRL.6]; 1986 edition tables 12 to 17) provide data on numbers of personnel involved in technical co-operation, either as experts or students and trainees.

9.8.2 *Ways of Categorising Manpower Data*

A broad summary of manpower (personnel overseas) is given in table 12 ([QRL.6]; 1986 edition). Four ways of categorising manpower are provided in this table:
(1) Numbers *recruited* during the year, to work overseas.
(2) Numbers *in post* at the end of the year.
(3) Numbers *financed* during the year; this combining the stock of personnel employed at the beginning of the year, plus additions during the year. The figure thus covers the number of individuals who have participated during the year, but gives equal weight to people involved for different periods of time (the whole year, or only part of it).
(4) Numbers of *man-years*. This overcomes the problem noted in the previous measure, by giving due weight to long-term and short-term assignments.

9.8.3 *Status of Manpower by Means of Financing*

The numbers covered in each of these alternative categorisations are divided into:
(1) *wholly-financed experts* employed by the UK Government, which meets all their emoluments and costs of employment. The part of this total accounted for by

technical co-operation officers is also given,

(2) *partly-financed experts* usually employed by overseas governments, but receive an inducement allowance from the UK Government in addition to their local salary. The part of this total accounted for by Overseas Service Aid Scheme (OSAS), and British Expatriates Supplementation Scheme (BESS), is also given,

(3) *volunteers* those who receive only an allowance from the government of the country in which they work, with the cost of their fares and the administrative costs of their sponsoring agency supported by the British Government.

9.8.4 *Technical Assistance Manpower by Assisted Region*

Data on UK technical assistance manpower (divided into wholly-financed, partly-financed and volunteers), by assisted region, is given in table 13 ([QRL.6]; 1986 edition). The data are expressed in terms of man-years (see section 9.8.2).

9.8.5 *Technical Assistance Manpower by Field of Assignment*

UK technical assistance manpower (divided into wholly-financed, partly-financed and volunteers, with additional sub-divisions of the first two categories), by field of assignment in 1985, is given in table 14 ([QRL.6]; 1986 edition). Twenty fields of assignment are covered. The data are expressed in terms of man-years (see section 9.8.2).

9.8.6 *Provision of Training in Technical Co-operation*

In *British Aid Statistics* ([QRL.6]; 1986 edition) table 15 provides a summary of the training component of TC, complementary to the manpower data of table 12 (see section 9.8.2). Four alternative ways of categorising training are provided, these being analogous to those introduced (in section 9.8.2) for manpower:

(1) Numbers of *new awards* during the year.
(2) Numbers *on course* at the end of the year
(3) Numbers *financed* during the year
(4) Numbers of *man-years* of training provided during the year.

The totals in each category are divided between those in Britain and those in their own or third countries. Further, each of these categories is divided between students and trainees. 'Students' pursue regular courses of study leading to an academic qualification, at a university or similar institute of higher education. 'Trainees' are normally in Britain only for a short time under arrangements connected with their existing employment. For 1985 only certain data compilation problems occured with respect to students and trainees. In table 15 this prevented the breakdown of totals between students and trainees.

9.8.7 *Training by Region*

Training is broken down by region in table 16 ([QRL.6]; 1986 edition). For 1983 and 1984 the division between students and trainees is also given, though as noted (see section 9.8.6) this could not be provided for 1985. These data are expressed in man-years.

9.8.8 *Training by Subject and Economic Sector*
Training in 1985 is divided by subject in table 17 ([QRL.6]; 1986 edition). 'Student' and 'trainee' statistics are originally compiled by subject being studied. When possible these subjects are allocated to an economic sector, the first half of the table providing training data on fifteen economic sectors. However, in the case of subjects which can benefit a range of sectors, and thus cannot be readily allocated to a specific sector (e.g. mathematics, electrical engineering, accountancy), a broader categorisation is used. Students and trainees studying such subjects are, in the second part of the table, allocated to either (i) natural sciences (ii) general engineering (iii) arts (iv) social sciences (v) unknown.

Once again the 1985 version of this table provides less information than normal on the breakdown between students and trainees (see section 9.8.6). Also, whereas the data normally takes the form of man-years, for 1985 only it is presented as 'people on course at year-end'.

9.8.9 *Technical Assistance Manpower and Training by Recipient Country and Sector*
Information on manpower and training in 1985, with a detailed recipient country breakdown, is provided in table 19 ([QRL.6]; 1986 edition). The total of man-years of experts and volunteers provided to each country is divided by sector. The sectors delineated are: (i) renewable natural resources (ii) education (iii) health (iv) other. For years other than 1985 (see section 9.8.6) similar information is presented on students and trainees, in terms of man-years of training provided.

The same table ([QRL.6]; 1986 edition; table 19) also provides, for each country, the numbers of experts and volunteers in posts at the year-end. The total is divided into: (i) wholly-financed (ii) partly-financed (iii) volunteers. Data are also given on the number of students and trainees on course at the year-end. For years other than 1985 the figures are sub-divided between students and trainees.

9.9 UK Foreign Aid—Comparative Performance

9.9.1 *Introduction*
Considerable interest exists on the comparative aid provision of major donor countries. Several compilations made available by international organisations present data on UK foreign aid performance relative to other sources.

9.9.2 *World Development Report Data*

9.9.2.1
A useful succinct source of data on UK aid provisions, placed in direct comparison with other leading donors, is the annual *World Development Report* of the IBRD [QRL.66]. In the detailed 'World Development Indicators' annexe to each annual report a table (e.g. table 21 in [QRL.66]; 1987 edition) provides information on aid flows. Four time series of data can be derived from these tables.

9.9.2.2

The first time series available from this source covers annual amounts of official development assistance (see section 9.2.5) in millions of US dollars. The amounts reported are net disbursements to developing countries and multilateral institutions. From the 1980 edition onwards the disbursements to multilateral institutions are reported for all members of the OECD Development Assistance Committee (DAC) on the basis of the date of issue of notes; previously, some DAC members having reported on the basis of the date of encashment.

Two sub-series are, in fact, available of this type. This first covers the ODA of UK and other OECD members. The tables permit a series for OECD members covering 1960, 1965 and annually from 1970 onwards. The second sub-series covers the ODA of OPEC members. The series first appeared in the 1979 edition of *World Development Report* [QRL.66] and comprises annual data from 1975 onward.

9.9.2.3

The second time series permitting comparison of UK aid performance with that of other donor countries, which can be derived from the World Bank source [QRL.66], expresses the ODA flows introduced above (section 9.9.2.2) as a percentage of donor country's GNP.

9.9.2.4

The third time series incorporated in the *World Development Report* [QRL.66] presents the ODA of OECD countries (for the years reported in section 9.9.2.2) in their national currencies.

9.9.2.5

The final time series covers net bilateral flows of the UK and other OECD countries, to low-income economies, expressed as a percentage of donor GNP. Net bilateral flows exclude unallocated bilateral flows and all disbursements to multilateral institutions. This series first appeared in the 1980 edition, and permits coverage of the years 1960, 1965, 1970 and annually from 1975 onwards.

9.9.3 *OECD Geographical Distribution of Financial Flows Data*

9.9.3.1

The OECD publication *Geographical Distribution of Financial Flows to Developing Countries* [QRL.34] provides a broader perspective (both by incorporating extensive recipient country data, and in terms of the range of information) on UK aid performance relative to other donors.

The objective of the OECD volume is to present detail on the external financing of individual developing countries and groupings of countries. However, as the country and area tables provide data on the various types of finance divided by source countries (including UK) and agencies, the relative performance of UK emerges.

Tables are provided for over 100 individual developing countries, and for eight groups of countries. The groupings covered are:
(1) Least-developed countries (LLDCs), as listed by the UN
(2) India and China
(3) Other LDCs
(4) Lower middle-income countries (LMICs), i.e. those with a per capita GNP in 1983 between $700 and $1,300
(5) Upper middle-income countries (UMICs); with a per capita GNP exceeding $1,300
(6) Unallocated
(7) Overall total (i.e. 1 to 6)
(8) Low-income countries total (i.e. 1 to 3); with per capita income below $700.

9.9.3.2

Each recipient country or group table incorporates fourteen sub-tables which include data on the UK as a source of finance. These sub-tables provide information on the following:
(1) Total net receipts, i.e. net official development assistance (ODA), plus net other official flows (OOF), plus net private sector flows
(2) Total gross official development assistance (see section 9.2.5)
(3) Total net official development assistance (see section 9.2.5)
(4) Official development assistance commitments (see section 9.3.4.1)
(5) The grant component of ODA (see section 9.3.4.2)
(6) Technical co-operation grants (part of 5)
(7) Gross ODA loans
(8) Net ODA loans
(9) Gross other official flows (see section 9.2.5)
(10) Net other official flows (see section 9.2.5)
(11) Gross total official flows (i.e. 2 plus 9)
(12) Net total official flows (i.e. 3 plus 10)
(13) ODA loans commitments
(14) Grant element of ODA loan commitments

9.9.3.3

From this information it is possible to derive the UK share of each type of finance to a particular developing country, and also to compare the structure of UK official finance to a given recipient with that of other donor countries.

9.9.4 *OECD DAC Development Co-operation Data*

9.9.4.1

The most detailed, and wide-ranging, international source of information on UK aid efforts, compared with that of other major donors, is the annual *Development*

Co-operation report of the Development Assistance Committee (DAC) of the OECD [QRL.26]. In some areas, notably that relating to qualitative aspects of aid (terms, tying etc.), this source provides some perspectives not fully covered by UK domestic sources. The 'Statistical Annexe' to the DAC Report [QRL.26] provides the basic and more systematic coverage, but *ad hoc* text tables sometimes provide additional information relating to particular issues or aspects of aid. A particularly important example of this is the detailed survey of aid to Sub-Saharan Africa in the 1986 Report ([QRL.26]; Part 4).

Except where explicitly mentioned below it may be assumed that all the tables reviewed provide UK data alongside, and comparable with, that for other OECD donors. Attention may also be drawn to the fact that the 1985 Report constituted a special review of *Twenty-Five Years of Development Co-operation* and, as such, provides a longer-term overview of aid performance.

9.9.4.2

The first three tables in the DAC Report Statistical Annexe ([QRL.26]; 1986 Report) are described as 'Aid Profiles'.

9.9.4.3

The first 'Aid Profile' table provides basic information on ODA volume, for 1984/85 average ([QRL.26]; 1986 Report; Table 1). Net disbursements data are given for:

(1) ODA as percentage of GNP
(2) Each country's percentage share in the total ODA of the DAC members
(3) Annual average growth rate of ODA in real terms (1979/80 to 1984/85)
(4) Total resource flows as a percentage of GNP (includes private flows and other non-ODA)
(5) Private grants as a percentage of GNP

In the 1985 Report [QRL.26] text table III-1 provides a long-term perspective on the performance of DAC members in terms of some of these indicators. Similarly text tables III-2 and III-3 in the 1986 Report [QRL.26] trace changing performance over more recent periods. Text table III-4 in the 1986 Report provides extra data attempting to assess performance adjusted for price and exchange rate changes.

9.9.4.4

The next Statistical Annexe table ([QRL.26]; 1986 Report; table 2) incorporates average 1984/85 information on 'other burden-sharing indicators'. These indicators are:

(1) Grant equivalent as a percentage of GNP, i.e. grant disbursements plus grant element of ODA loans commitments applied to gross loan disbursements of ODA
(2) Multilateral ODA as a perecentage of GNP
(3) Aid to International Development Association (IDA), International Fund for Agricultural Development (IFAD) and UN Development Programme (UNDP) as a percentage of GNP

(4) Aid to Low Income Countries (LICs—i.e. 1983 per capita income below $700) as a percentage of GNP
(5) Aid to least-developed countries (LLDCs—i.e. the 40 countries listed by UN) as a percentage of GNP

Text table III-5 ([QRL.26]; 1986 Report) gives information on ODA appropriations as a percentage of central government budget expenditures by DAC members. The data are given as three year averages, for four selected periods, covering 1970 to 1985.

9.9.4.5

The third of the 'Aid Profiles' tables ([QRL.26]; 1986 Report; table 3) deals with qualitative aspects of ODA, and is sub-divided into four parts. In all cases the figures (1986 Report) are the 1984/85 average.

The first sub-section deals with *financial terms of ODA commitments* (excluding debt reorganisation), providing three figures:

(1) Grant element of total ODA
(2) Grant element of ODA to least developed countries
(3) ODA loans with grant element under 50 per cent as a percentage of bilateral ODA.

For discussion of further tables relating to terms of ODA see section 9.9.4.22 to 9.9.4.24. The second sub-section of table 3 refers to the *tying and local cost* status of aid, providing figures for:

(1) Untied aid as a percentage of ODA
(2) Bilateral untied aid as a percentage of bilateral ODA
(3) Local cost financing as a percentage of bilateral ODA, where local costs are outlays up to the date of completion of a project to finance the procurement of goods and services in the local market.

More information on tying is given in tables described in section 9.9.4.13 and 9.9.4.25. The table's third section covers five aspects of *multilateral assistance and geographical distribution*:

(1) Multilateral aid as a percentage of total ODA
(2) Aid to low income countries as a percentage of total ODA (i.e. bilateral ODA plus an imputed share of multilateral contributions)
(3) Aid to least developed countries as a percentage of total ODA (as in 2)
(4) Bilateral ODA to lower middle-income countries (LMICs—i.e. recipients with per capita income between $700 and $1300 in 1983) as a percentage of bilateral ODA
(5) Bilateral ODA to upper middle-income countries (UMICs—i.e. per capita income exceeding $1300 in 1983) as a percentage of bilateral ODA.

The final section of table 3 covers *sector priorities*, under five headings:

(1) Bilateral aid to agriculture (including rural infrastructure, regional and river basin development schemes, but excluding agro-industries) as a percentage of bilateral ODA (commitments), see also section 9.9.4.27 below
(2) Aid to agriculture (commitments) as a percentage of total ODA (including imputed multilateral flows)
(3) Aid to energy as a percentage of total ODA (commitments)

(4) Aid to education as a percentage of total ODA (commitments)
(5) Technical co-operation in developing countries (excluding expenditure for students and trainees in the donor countries and cultural programmes) as a percentage of bilateral ODA. This information is for average 1982/83 disbursements.

Further information on the division of ODA by sector or purpose is discussed in section 9.9.4.28.

9.9.4.6

The next group of tables in the DAC Report Statistical Annexe ([QRL.26]; 1986 Report) cover 'basic resource flows'. Three of these 'basic resource flow' tables incorporate UK data for comparison with other OECD countries. In each case the tables include average data for 1974/76 and annual figures for 1980 to 1985 (1986 Report), providing net disbursements (at current prices and exchange rates) data in $ million and as a percentage of GNP.

9.9.4.7

Table 5 ([QRL.26]; 1986 Report) covers the total net flow of resources from DAC countries to developing countries and multilateral agencies. This aggregate encompasses official development assistance, other official flows, private flows and grants by private voluntary agencies.

9.9.4.8

The next table ([QRL.26]; 1986 Report; table 6) covers the net flow of private capital from DAC countries to developing countries and multilateral agencies. This aggregate incorporates direct investment, bilateral portfolio investment, multilateral portfolio investment and export credits but excludes grants by voluntary agencies.

9.9.4.9

The last of this group of tables ([QRL.26]; 1986 Report; table 7) covers net official development assistance from DAC countries to developing countries and multilateral agencies. This aggregate incorporates bilateral grants and loans and contributions to multilateral institutions, but excludes other official flows (see section 9.2.5).

9.9.4.10

Two Statistical Annexe tables ([QRL.26]; 1986 Report) give details on individual DAC members' financing of multilateral agencies. In table 8 data are given for donor countries' net disbursements to various agencies in 1985. The agencies for which information is provided are:
(1) The World Bank Group; with separate figures for:
    (a) IBRD; including International Financing Corporation (IFC)
    (b) IDA
(2) Inter-American Development Bank (IDB); with separate figures for:

(a) ordinary capital
(b) special fund

(3) Asian Development Bank; with separate figures for:
(a) ordinary capital
(b) special fund
(4) African Development Fund; including contributions to the African Development Bank
(5) EEC; with figures for the part of such contributions going to the European Development Fund
(6) UN Agencies
(7) Other

9.9.4.11

Capital subscriptions by individual DAC members to multilateral agencies are reported in table 9 ([QRL.26]; 1986 report). Annual data are given for 1980 to 1985, on both the deposits and encashments basis.

9.9.4.12

Some longer-term perspectives (various periods between 1974 and 1985) on DAC members' contributions to multilateral development agencies and funds are reported in text table VI-2 in the 1986 DAC Report [QRL.26]. For each country such contributions are expressed as a percentage of: (i) its ODA (ii) its GNP (iii) total DAC multilateral ODA, excluding EEC.

9.9.4.13

Information on the tying status of DAC members ODA (gross disbursements) in 1985 ([QRL.26]; 1986 Report) is given in table 12. Bilateral ODA is divided into three categories:

(1) Untied, i.e. fully and freely available for essentially world-wide procurement. This untied component is further divided into:
(a) Cash, i.e. amounts not directly financing imports, such as budget and balance of payments support, local cost financing and debt relief,
(b) Grants
(2) Partially untied, i.e. contributions available for procurement from the donor and also from substantially all developing countries
(3) Tied, i.e. aid tied to procurement in the donor country. Figures are also given for the part of tied aid accounted for by technical co-operation. Multilateral ODA (excluding that through the EEC) is sub-divided as tied or untied.

For further data on tying see section 9.9.4.25.

9.9.4.14

It should be noted that at the time of compilation of the 1986 Report the concepts and definitions used in the table described above were under review. Under possible reformulated definitions some formally untied aid (particularly technical assistance) would be classified as "in effect" tied.

9.9.4.15

Three tables in the DAC Report Statistical Annexe ([QRL.26]; 1986 Report) deal with aspects of technical co-operation:

(1) Technical co-operation expenditure, including technical cooperation loans (table 14). The information is for official bilateral disbursements by individual donors.

(2) Numbers of students and trainees (table 15). The figures are for publicly financed fellowships by individual donors.

(3) Numbers of experts and volunteers (table 16). The figures are for publicly financed technical co-operation personnel by individual donors.

In each of the tables information is given for 1970 and annually for 1980 to 1985.

9.9.4.16

In the 1986 Report [QRL.26] table 18 provides listings, for each major donor country, of the leading recipients of its ODA. Each leading recipient country's percentage share of the donors total ODA (gross disbursements) is given. The UK section of this table provides rankings for 1970/71, 1980/81 and 1983/4.

9.9.4.17

The major integrated summary of the UK's flow of financial resources (including private) to developing countries and multilateral agencies, provided in the DAC Report ([QRL.26]; 1986 Report), is contained in table 35. This table provides information on the 1974/76 average and annual figures from 1982 to 1985. Identically-presented coverage is provided for the other DAC donor countries. To facilitate direct inter-country comparisons information is provided for 1984 (table 28) and 1985 (table 29) by type of flow and DAC donor. In the main parts of these tables, outlined below, the data are given for net disbursements. However, less detailed information is also given for gross disbursements and commitments.

9.9.4.18

In the tables introduced in section 9.9.4.17 above the first major component covered is *official development assistance* (ODA). This is sub-divided firstly under (1) bilateral ODA and (2) contributions to multilateral agencies as follows:

(1) *Bilateral ODA*, divided into:

(a) grants and grant-like contributions; comprising:
- (i) technical assistance
- (ii) food aid
- (iii) administrative costs
- (iv) other grants

(b) development lending and capital; comprising
- (i) new development lending
- (ii) food aid loans

(iii) equities and other bilateral assets.

(2) *Contributions to multilateral institutions* are divided into:

(a) grants to:
  (i) UN agencies
  (ii) EEC
  (iii) other

Data are also given on the total food aid grants incorporated in the three components (though not for the food aid part of each component).

(b) Capital subscription payments and similar to:
  (i) IBRD (including IFC)
  (ii) IDA
  (iii) regional development banks
  (iv) other

(c) concessional lending.

A memo item later in the table reports interest received on ODA.

9.9.4.19

The second major component of these summary tables is *other official flows* (OOF). This is sub-divided into (1) bilateral OOF and (2) multilateral institutions. Bilateral OOF is further divided into (a) official export credits and (b) equities and other bilateral assets. Memo items later in the table provide data on (i) official funds in support of private export credits, which is a part of (a), and (ii) official funds in support of private investment, which is a part of (b). A further memo item reports interest received on OOF.

9.9.4.20

ODA and OOF comprise *total official flows* in the summary tables under review (see section 9.9.4.17). The contribution of private capital to total resource flows to developing countries and multilateral agencies is also divided into two parts i.e. *grants by private voluntary agencies* (with no sub-division of information) and *private flows at market terms.*

9.9.4.21

Four components are provided for *private flows at market terms*:

(a) direct investment
(b) bilateral portfolio investment and other
(c) multilateral portfolio investment
(d) private export credits

In addition data are given on two contributions of the monetary sector to private flows, which are encompassed in the four components listed above. These are (i) change in bilateral claims of resident banks (including affiliates in the UK of banks with headquarters elsewhere), (ii) multilateral portfolio investment of monetary sector.

9.9.4.22

The financial terms and similar conditions relating to aid flows are a source of considerable concern. The DAC Reports' more detailed data on these characteristics tend to be incorporated in text tables. Table III-9 in the 1986 Report, for example, provides an indication of trends in financial terms of DAC members' ODA commitments. Data are provided on (a) grant element of total ODA, including debt reorganisation (various periods from 1970 to 1985) and (b) grant element of ODA to LLDCs (various periods from 1975 to 1985).

9.9.4.23

Similarly, text table III-10 in the 1986 Report provides more detail on ODA terms (excluding debt reorganisation) for 1984 and 1985. This covers (a) grants as a share of ODA commitments (including grants and capital subscriptions to multilateral agencies), and (b) information on four characteristics of ODA terms, i.e. (i) interest rate, (ii) maturity (years), (iii) grace period (years), (iv) grant element.

9.9.4.24

Text table III-11 in the 1986 DAC Report evaluates members compliance, in 1984 and 1985, with the 1978 DAC terms recommendation. These 1978 terms recommendations are described in the 1986 Report ([QRL.26]; 1986 Report; pp. 59-60). For comparison with these targets, data are provided on:

(1) ODA commitments (excluding debt reorganisation); $ million
(2) Grant element of ODA commitments; percentage
(3) Grant element of ODA commitments to LLDCs; percentage
(4) ODA commitments as a percentage of GNP.

9.9.4.25

Another aspect of the conditions relating to aid, on which extra information is available in text tables, is the extent of tying. In the 1986 DAC Report [QRL.26] text table III-12 provides annual figures (1983 to 1985) for bilateral tied and partially untied ODA commitments (for these concepts see section 9.9.4.13) and for the part of this which had a less than 50 per cent grant element. Text table III-13 expresses the information of table III-12 as a percentage of total bilateral ODA commitments.

9.9.4.26

A concept with some similarities to the more traditional aid tying, and whose recent emergence has stimulated similar concerns and debates, is that of Associated Financing (also known as 'mixed credits'). Associated Financing represents the combination of official development assistance with export credits, and is discussed in some detail in the 1985 DAC Report ([QRL.26]; 1985 Report; pp. 243-247). In the 1986 Report text table III-12 provides annual figures (1983 to 1985) on the total value of associated financing (AF), and that part accounted for by aid funds (AF-ODA). Text table III-13 ([QRL.26]; 1986 Report) expresses AF-ODA as a percentage of total bilateral ODA commitments for 1983 to 1985.

9.9.4.27

Another aspect of aid which has been given special coverage in text tables is food and agriculture. Thus in the 1986 Report [QRL.26] table IX-2 gives annual (1978-84) official development commitments to food and agriculture, whilst table IX-3 expresses these as a share of total ODA commitments. Table IX-4 breaks down official development financing commitments to food and agriculture into fourteen sub-sectors, for 1984.

9.9.4.28

The 1985 DAC Report [QRL.26] provided (in statistical annexe table 7) information on the allocation of bilateral ODA commitments by ten broad sectors (health, education, agriculture etc.). The data covers that part of bilateral ODA commitments which it is possible to allocate by sector in this way. In table 6 ([QRL.26]; 1985 Report) bilateral ODA commitments not allocatable by sector, as above, are divided by purpose. The purposes delineated are:

(1) contributions to finance current imports—with data on the part accounted for by food aid
(2) emergency and disaster relief other than food aid
(3) contributions not linked to imports
(4) debt reorganisation
(5) other bilateral aid not allocated by sector.

Though available from several earlier DAC Reports, the 1986 edition did not provide this information.

CHAPTER 10

INTERNATIONAL FLOWS OF LABOUR OF THE UNITED KINGDOM

10.1 Basic information

In contrast to international flows of capital and, to a lesser extent, technology, flows of labour receive relatively little attention. The basic source of information on international migration is the Office of Population Censuses and Surveys *International Migration* [QRL.38]. This annual series (OPCS Series MN of which 1985 is No.12) covers migrants entering or leaving the United Kingdom with separate data for England and Wales. The last ten years are given together with the detailed figures for the most recent year: the 1987 edition covers 1985 in detail.

International Migration [QRL.38] covers:

(1) The analysis of population change. Estimates of UK net migration, which are used to estimate the change in the size of the population of the country as a whole, are shown for a ten year past period. Information on natural change is also given, as is total migration for each of the constituent countries of the UK.

(2) Time series tables for the UK and England and Wales separately are given [QRL.38]. These data are derived from the International Passenger Survey (IPS). It should be noted that figures under (1) above are derived from IPS figures which have been adjusted for some years in the light of other information and the figures in the two sections are not directly comparable.

(3) A detailed cross-tabulation of international migration for the most recent year is given by key variables for the United Kingdom.

(4) Home Office statistics are given for the past four years on (i) acceptances for settlement on arrival and on removal of time limit and (ii) people acquiring British citizenship, together with country of former citizenship. These figures are a by-product of the administration of the Immigration Acts. Definitions of settlement should be checked carefully when using these data. The coverage by country also differs from tables derived from IPS. IPS figures also include various categories which Home Office figures do not cover, e.g. British citizens. A detailed study of differences is given in [B.40]. Selected results from recent British Censuses of Population can be found in [B.42].

10.2 International Passenger Survey (IPS)

The IPS is a voluntary sample survey conducted by OPCS which covers the principal air and sea routes between the UK and overseas but excludes those between the UK and the Irish Republic (see also 10.3). It became fully operational in 1964. The survey samples between 0.1 per cent and four per cent of passengers, a proportion which varies according to route and time of year.

Internationally agreed definitions of migrants are used. A migrant into the UK is a person who has resided abroad for a year or more and, on entering, has declared the intention to stay in the UK for a year or more. A migrant from the UK is a person who has resided in the UK for a year or more, who on leaving, has declared the intention to reside abroad for a year or more. Travellers who are migrants are asked a group of questions which form the basis of the IPS migration statistics. These definitions do *not* coincide with the legal definitions contained in the various UK Immigration Acts (and therefore with the Home Office statistics). Figures obtained from the IPS are subject to sampling and non-sampling errors which make the estimates for small categories unreliable (see [QRL.38] No.10, 1983 pp. viii to xii). The figures are based on *intentions* to migrate which may not be realised.

A re-weighting of the sample in 1975 (see [QRL.38] No.2, 1975) means that figures from 1975 are not directly comparable with earlier years. Methodological revisions took place in 1976 (see [QRL.38] No.3, 1976 pp. 8-10).

The 1985 IPS questionnaires for air passengers are shown in the volume [QRL.38] (No.12, 1985). The questionnaires for sea passengers differ very little from those for air passengers.

10.3 International Migration

Ten time series tables are presented in [QRL.38]. These relate to citizenship (defined as the nationality of passport which the traveller is carrying). The British Nationality Act of 1981 (in force 1 January 1983) replaced citizenship of the UK and Colonies with British citizenship, British Overseas citizenship and British Dependent Territories citizenship. In the 1985 volume 'United Kingdom citizenship' (used in volumes prior to 1983) has been replaced by 'British citizenship' which includes British Overseas citizenship and is virtually equivalent to the old category. Persons with British Dependent Territories citizenship are shown under the country which issued their passport. Definitions of areas and political groupings should be checked carefully, particularly where time series are being compiled. Time series tables cover country of last or next residence with inflow, outflow and annual balance for each country or area. Country of birth is given in a separate series. A time series of international migration by occupational category is given by the following categories: professional and managerial (including administrators, managers and people with professional or technological qualifications), manual and clerical (all occupations excluding professional and managerial), students, housewives, others (including retired people, people with no occupation etc.) and children under 16. A time series profile by age and by sex and marital status is also given. International migration is also subdivided by area of destination or origin within the UK (countries of the UK and standard regions of England) and inflow of migrants to the UK by selected country of last residence.

The detailed cross-section of international migration for the latest year cross-classifies the data by the key variables mentioned above.

Some quarterly data on international migration are published routinely in *Population Trends* [QRL.51]. Data for the most recent four quarters are published twice yearly in *OPCS Monitors* [QRL.47]. Occasional articles on international migration occur in *Population Trends* [QRL.51] and in Occasional Papers published by the Office of Population Censuses and Surveys.

CHAPTER 11

DESIRABLE IMPROVEMENTS AND FUTURE DEVELOPMENTS

11.1 General Problems

In surveying the statistics of the international aspects of UK economic activities, several key problems emerge. First, there is the problem of the incompatibility of different national treatments of the same economic aggregate. For instance, US figures on UK direct investment in the USA differ from British accounts of the same figure. Partly, this is due to definitional differences but partly it is due to the accuracy of measurement and coverage of different national authorities. These international discrepancies are in fact growing as the analysis of the world current account shows (pages 21–22 above). Problems of this kind can only be dealt with by increasing international cooperation and exchange of information. Liaison between national statistical offices is, to say the least, rudimentary. Bodies such as OECD, IMF and GATT are making efforts on coordination but even basic definitions, such as the proportion of voting equity which is taken to mean 'control' of the enterprise located overseas, are not fully standardised.

The second problem is that the international scene is changing very rapidly and national statistics authorities are having great difficulty in keeping up with the pace of change. This is particularly true in respect of invisible transactions, especially in the area of 'new financial instruments'. Given the vast amounts of capital which flow across national boundaries, it is vital that records are as accurate as possible. There needs to be careful measuring of such flows for taxation purposes and the demise of Exchange Control has meant that information in this area is woefully inadequate and very dated.

A third major change affecting Britain, in particular, is the imminence of the 'Single European Act' which, in 1992, will create a unified market across the European Communities. The virtual disappearance of customs barriers within the EC will mean far less direct measurement of trade flows and the removal of tariffs generally has a detrimental side-effect on the recording of trade.

Fourth, and pervading all the above elements, is the decline in the resources available for statistical collection and compilation of information. Successive series of governmental cut-backs on the CSO, HM Customs and Excise and other bodies have had the inevitable result of a reduction in the quantity and quality of data.

Fifth, there is a need to increase the availability of machine-readable data. This volume has pointed to areas where this is available but ease of access to these data could improve and the scope of machine-readability could be widened.

11.2 The UK Balance of Payments

The CSO 'Pink Book' [QRL.61] will be the starting point for most users interested in international aspects of UK economic activities. As Chapter 1 showed, the quality of the information available in the Pink Book is very variable. Investigation of the minutiae of collection and coverage is vital for users. The quantitative grading of errors in the aggregates, abandoned in 1976, should be re-instated as this was a major help to users. Indications of the likely direction of errors would also be a useful addition. Refinement and updating of the analyses of reliability given in Tables 1.2–1.4 would be a major step forward.

The abolition of Exchange Control in 1979 meant the discontinuation of many statistical series which underlie the elements of the accounts. Many of these series have been merely trended forward using the nearest available series. The abandonment of quantitative estimates of individual items (as shown in Table 1.2) is a retrograde step.

11.3 UK Visible Trade

A basic problem with UK visible trade data is the poor quality of the basic returns by importers and exporters: a high proportion of customs documents have been shown to have significant errors on them. As tariff barriers decline, measurement errors are likely to increase (goods flows across frontiers where tariffs are in place are more accurately measued than custom-free boundaries). To maintain, let alone improve, the quality of information, a high degree of sampling measurements must be carried out as a check on accuracy.

In using international trade data, especially when combining results with domestic data, it is often necessary to assume a correspondence between categories SITC and SIC classifications. Trade data from SITC codes are transformed into SIC to produce [QRL.15] *Business Monitor MQ10*. Publication of an authoritative concordance would help researchers and other users enormously.

In the field of visible trade, as elsewhere in this volume, index number problems (choice of base year, weighting etc) arise. In visible trade statistics revaluations are made for twenty per cent of goods by taking prices of a similar commodity or group of commodities, with a further ten per cent 'guesstimated'. Clearly, more direct measurement of volume and price would greatly improve accuracy in this crucial area (see section 2.10 above).

11.4 UK Invisible Trade

The nature of invisible trade is constantly changing and this gives rise to huge problems in both definition and measurement. Definitional problems arise particularly with regard to new financial instruments. Table 1.2 shows a dated estimate of the likely reliability as of 1976 and this reliability has declined in many areas consequent on the abandonment of Exchange Control and the reduction in direct recording.

11.4.1 *Services*

Forty per cent of service credits are based on annual data collected annually or less frequently collected data or no directly reported data. Particular problems occur in 'other services' which is a catch-all category.

Information on travel is dependent on the International Passenger Survey. Clearly, an increase in the number of passengers questioned (and responses are voluntary) would improve the results—in 1985 0.24 per cent of travellers were surveyed. This, however, has costs. Error levels are given in section 3.2.5.1 above. Here it is sufficient to note that sampling errors vary widely according to type of travel and route. A major improvement with tourism into the UK would be an improvement in regional information. This is collected by region on an annual basis but is not given quarterly nor monthly and regional aspects are not given the prominence which they deserve because of the very broad categories used (London, 'Other England', Scotland and Wales). The increasing importance of the tourism industry, particularly to the regions, seems to require an intensification of effort in the collection of tourism statistics.

Financial services suffer particularly from definitional problems, lack of direct measurement and the dynamic nature of the industry. Efforts to measure the impact of new financial instruments naturally lag behind their introduction. The basic source of much information is the Bank of England and directly from the active institutions themselves. However, many surveys are carried out irregularly and many of the series are now seriously dated and are merely trended forward. A fairly wide survey of this area is now overdue in order to cast the statistical net more widely and to capture new developments.

11.4.2 *Earnings of UK External Assets and Liabilities*

As elsewhere, a major difficulty with the recording of interest, profits and dividends (IPD) is that the information comes largely from the companies themselves. The companies' own estimates of depreciation are used and these are, in the main, entered at historic costs. Attempts to move to current costs are likely to be difficult and expensive but it would improve the quality of the information.

11.4.3 *'The City'*

The presentation of the earnings of 'The City' separately in the 'Pink Book' [QRL.61] has led to criticism, particularly with regard to the recording of net earnings. This is unobjectionable as long as the meaning of net earnings is made clear and users are aware of the problems of presentation (see section 3.4 above).

11.4.4 *Transfers*

Statistics of international private transfers are extremely unreliable and are likely to remain so given the nature of the unrequited transactions which they cover. The quality has deteriorated since the end of Exchange Control and can only be improved by spot sample checks. However, this may not be an effective use of resources given problems of greater magnitude elsewhere. General government transfers will generally be accurate.

11.5 International Capital Movements

The increasing importance of international capital flows and their changing nature repesent a major challenge for statistical offices. In particular, the recording of levels of UK external assets and liabilities at book values (for direct investment) means that estimates of the inward and outward stock of direct investment rapidly becomes dated.

11.6 Outward and Inward Direct Investment

Information on the UK direct investment position derives largely from the DTI surveys, discussed in chapter 5. The annual series records flows of new direct investment and earnings. The theoretically triennial survey compiles data on stocks at given moments of time.

The stock survey collects book values rather than the conceptually more satisfactory market or replacement cost data, but as this information is provided by the firms themselves, this is regarded as being the only feasible basis for collection. It is important to make the points that this series has slipped from time to time from its triennial norm (see Appendix 5) and the 1984 enquiry was greatly delayed. Secondly, although the DTI inquiry (since 1978) has been statutory for outward investment, it remains voluntary for inward investment. Thirdly, there seems to have been an accelerating trend over recent years to reduce the requirements in the questionnaire to companies, no doubt reflecting the Conservative government's general attitude to reducing the burdens of the state on private enterprise. However, this has the unfortunate side-effect of reducing the information available to those companies and, as we know, companies require information in order to function effectively. Fourthly, between surveys, interpolated information is used, the accuracy of which cannot be guaranteed. Fifthly, a major omission is information on *sales* of goods and services of foreign subsidiaries of UK companies and of foreign companies in the UK. This omission is a vital one, particularly in view of the fact that the US Department of Commerce *Survey of Current Business* does produce such sales figures of US subsidiaries abroad (by industry). Collection problems are not, therefore, insuperable.

The annual transactions survey uses a definition of direct investment at variance with the OECD guidlines (twenty per cent of equity rather than ten per cent) and this does not aid international comparison. Figures on gross direct investment and disinvestment have been dropped since the 1981 enquiry, only net figures now being given. The separate amounts of investment and disinvestment are thus lost. Disinvestment is an important subject in its own right and its investigation is now seriously hampered.

11.7 Royalties and Miscellaneous Transactions

The increasing importance of issues centered on the 'technological balance of payments' makes data on royalties important in the debate on growth and competitiveness. Royalty payments are reported from 'related' and 'unrelated' concerns (the former having an ownership link) and these are used as indicators of technology flows. But in this use they must be treated with great care as it is difficult to attribute prices to flows between 'related' concerns. The reported plan for a benchmark inquiry into overseas payments for royalties, patents and licenses (by DTI) is greatly to be welcomed. A comprehensive benchmark survey would enable the series to be anchored and better projections can then be drawn of overall royalty payments and receipts. The internal pricing problems will, however, remain.

The provisions of the Companies Act 1981 had the unfortunate result of ending the series on direct exports as from that year, this represents an irreplaceable loss of valuable information.

11.8 International Transactions of the UK Monetary and Banking Sector

The task of statisticians in keeping up with changes in the burgeoning monetary and banking sector has been noted elsewhere. The use of data drawn largely from balance sheets allows consistency but occasionally obscures meaning. Occasional difficulties in securing long runs of data are encountered, often as a result of changes in banking practice.

11.9 Sterling Exchange Rates

A useful step forward in recent years has been the derivation of composite exchange rate indices (see section 8.1.2). Users should, however, note that these *are* artificial constructions and recognise the need to understand the nature of their derivation.

11.10 UK Foreign Aid

It's clear that an understanding of the adequacy of efforts in the area of foreign aid needs to take account of many dimensions beyond the purely financial. It would seem that, in the main, the figures provided by Overseas Development Administration achieve commendable recognition of this.

11.11 International Flows of Labour of the UK

International flows of labour have received little attention compared to capital flows but they are likely to be increasingly important, particularly within the projected Single European Market. The basic source of information derives from the voluntary International Passenger Survey. A finer breakdown by occupational groups would now appear to be necessary for manpower planning and training policies. But calls for further information in these areas must be balanced by the conflicting need for personal privacy.

11.12 Summary

International trade, finance and investment, by definition, cross national boundaries and jurisdictions. The aim of statisticians in this area must be that both ends of the transactions are similarly recorded in the different nations. This has not yet been achieved with any degree of symmetry—indeed, in many areas there has been a decline in equal and opposite recording of data. Not only is greater international co-operation necessary but increasing resources need to be devoted to the accurate recording and analysis of trade and investment data. General policies of liberalisation have had the unfortunate side-effect of reducing recording mechanisms (exchange control records, tariff records). In some areas the active institutions, traders, financiers, and investors have moved far too quickly for statistical records to encompass their activities, notably in the new financial markets. It should not be forgotten that markets require information in order to function effectively and the provision of this information, whilst having undoubted costs, is a major contribution to effective functioning.

QUICK REFERENCE LIST TABLE OF CONTENTS

QUICK REFERENCE LIST

Type of data	Breakdown	Area	Frequency	QRL publication	Text reference
UK Direct Investment					
Book values of assets	By area and country, i.e. country of origin for inward investment, host country for outward investment	UK	Mainly triennial	[QRL.10]	5.2.2 for outward investment. 5.4.2 for inward investment. For frequency see 5.2.1.1
Book values of assets	By industry, outward and inward investment	UK	Mainly triennial	[QRL.10], [QRL.5]	5.2.3 for outward investment. 5.4.3 for inward investment. For frequency see 5.2.1.1. Also see Appendix 5
Book values of assets	By form of organisation, outward and inward investments	UK	Mainly triennial	[QRL.10], [QRL.5]	5.2.4 for outward investment. 5.4.4 for inward investment. For frequency see 5.2.1.1. Also see Appendix 5
Book values of assets	By percentage shareholding, outward and inward investments	UK	Mainly triennial	[QRL.10], [QRL.5]	5.2.5 for outward investment. 5.4.5 for inward investment. For frequency see 5.2.1.1. See also Appendix 5
Book values of assets	By size of enterprise, outward and inward investments	UK	Mainly triennial since 1971	[QRL.10]	5.2.6 for outward investment. 5.4.6 for inward investment. For frequency see 5.2.1.1

Book values of assets	Balance sheet analysis of assets and liabilities, outward and inward investments	UK	Mainly triennial	[QRL.10], [QRL.5]	5.2.7 for outward investment subsidiaries. 5.2.8 for outward investment branches. 5.4.7 for inward investment subsidiaries. 5.4.8 for inward investment branches. For frequency see 5.2.1.1. See also Appendix 5
Annual investment flows	By area and country: i.e. country of origin for inward investment, host country for outward investment	UK	Annual since 1960	[QRL.10], [QRL.5]	5.3.2 for outward investment. 5.5.2 for inward investment. See also 5.3.1.7
Annual investment flows	By industry and country, outward and inward investment	UK	Annual since 1965	[QRL.10], [QRL.5]	5.3.3 for outward investment. 5.5.3 for inward investment. See also 5.3.1.7
Annual investment flows	By components of investment, outward and inward investment	UK	Annual since 1960	[QRL.10]	5.3.4 for outward investment. 5.5.4 for inward investment. See also 5.3.4.2 and 5.5.4.2
Annual earnings	By area and country: i.e. country of origin for inward investment, host country for outward investment	UK	Annual since 1960	[QRL.10], [QRL.5]	5.3.5 for outward investment. 5.5.5 for inward investment. See also 5.3.1.7
Annual earnings	By industry and country, outward and inward investment	UK	Annual	[QRL.10], [QRL.5]	5.3.6 for outward investment since 1965. 5.5.6 for inward investment since 1973. See also 5.3.1.7
Annual earnings	By components of earnings, outward and inward investment	UK	Annual since 1960	[QRL.10]	5.3.7 for outward investment. 5.5.7 for inward investment. See also 5.3.7.1 and 5.5.7.1

Type of data	Breakdown	Area	Frequency	QRL publication	Text reference
Net receipts from overseas affiliates	Flows of dividends, interest and net payments for royalties and services from overseas affiliates	UK	Annual since 1960	[QRL.10], [QRL.5]	5.3.8. Differs from 'earnings' by excluding unremitted earnings and including royalties and service fees etc. See also 5.3.1.7
Net payments from UK affiliates to overseas investors	Flows of dividends, interest and net payments for royalties and services from UK affiliates to overseas investors	UK	Annual since 1960	[QRL.10], [QRL.5]	5.5.8. Differs from 'earnings' by excluding unremitted earnings and including royalties and service fees etc.
Foreign enterprises in UK manufacturing industry, Census of Production data	By country of origin	UK	Biennial since 1973 plus 1984	[QRL.17]	5.6.2. Includes data on variables (e.g. employment, sales, wages) not covered by DTI Direct Investment Surveys
Foreign enterprises in UK manufacturing industry, Census of Production data	By industry	UK	Biennial since 1973 plus 1984	[QRL.17]	5.6.3
Foreign enterprises in UK manufacturing industry, Census of Production data	By UK country and region	UK	Biennial since 1973 plus 1984	[QRL.17]	5.6.4
Foreign enterprises in UK manufacturing industry	Participation of foreign enterprises in the 100 largest UK private sector manufacturing enterprises	UK	Annual since 1973	[QRL.17]	5.6.5

Royalties etc.					
Transactions, aggregate values, information on receipts, expenditure and balance	Divides royalties into: (1) artistic royalties, (2) technological and mineral royalties. Also divides transactions into those with: (1) related concerns overseas, (2) unrelated concerns overseas	UK	Annual since 1964	[QRL.10]	6.2.2 Includes estimates for non- respondents to surveys. See also Appendix 8
Transactions, receipts and expenditures	By area and country. Also division by 'type' and 'related-unrelated' (see above)	UK	Annual since 1964	[QRL.10]	6.2.3. Only covers 'returns received' in survey. See also Appendix 8
Transactions, receipts and expenditures	By industry. Also division between 'related' and 'unrelated' (see above)	UK	Annual since 1964	[QRL.10]	6.2.4 only covers 'returns received' in survey. Frequency changes in industry coverage, see 6.2.4.2. See also Appendix 8
Film and Television Transactions					
Overseas transactions in film and TV material. Receipts and expenditures	Summary table providing separate data on: (1) film companies, (2) TV companies	UK	Annual since 1963	[QRL.10]	6.3.2. For frequency see 6.3.2.2. See also Appendix 9
Overseas transactions in cinema films. Receipts and expenditures	By geographical area. By type of transaction: cinema and TV, film	UK	Annual since 1964	[QRL.10]	6.3.3. See also Appendix 9
Overseas transactions in cinema films, receipts and expenditures	By type of transaction i.e. 'performance' or 'production'. By type of company	UK	Annual since 1965	[QRL.10]	6.3.4. See also Appendix 9
Overseas transactions of BBC and IBA programme contractors. Receipts and expenditures	By type of transaction i.e. 'performance' or 'production'. By geographical area	UK	Annual since 1975	[QRL.10]	6.3.5. For frequency see 6.3.5.2. See also Appendix 9

Type of data	Breakdown	Area	Frequency	QRL publication	Text reference
International Transactions of UK Monetary and Banking Sector					
Banks balance sheets, monthly reporting institutions	Information provided for seven groups of monthly reporting institutions i.e. type of bank	UK	Monthly	[QRL.4]	7.2.7
Balance sheets, quarterly reporting institutions		UK	Quarterly since third quarter 1986	[QRL.4]	7.2.3
Balance sheets, discount market		UK	Monthly	[QRL.4]	7.2.4
Consolidated balance sheet of UK monetary sector	Shows position of UK monetary sector as a whole in relation to third parties	UK	Quarterly	[QRL.4]	7.2.5
Overseas contribution to financing central government borrowing requirements (CGBR)	Divides CGBR into, (i) external and foreign currency finance, (ii) domestic sterling finance provides data on components of (i)	UK	Quarterly	[QRL.4]	7.3.1
External transactions in marketable government debt		UK	Quarterly	[QRL.4]	7.3.2
External liabilities and claims of UK monetary sector and other financial institutions		UK	Quarterly	[QRL.4]	7.4
Consolidated external claims and unused commitments of UK registered monetary sector institutions and their worldwide branches and subsidiaries	Detailed country coverage	UK	Quarterly	[QRL.4]	7.5

External liabilities and claims in sterling reported by UK monetary sector and other financial institutions	Data provided on five components of sterling claims and liabilities. Area breakdown is provided for each component	UK	Quarterly	[QRL.4]	7.6
UK official reserves	Divides reserves into four components	UK	Monthly	[QRL.4]	7.7
UK government and other public sector foreign currency debt	Distinguishes, (i) official debt, (ii) public sector commercial debt. Also components of each of these	UK	Monthly	[QRL.4]	7.8
Direct and portfolio investment of banking sector		UK	Annual	[QRL.61]	7.9.2; 7.9.3; 7.9.4
Sterling Exchange Rates					
Sterling exchange rate index	Sterling ERI based on 1975 = 100	UK	Monthly average	[QRL.33], [QRL.30]	8.1.2 (definitions) 8.2.1.1 8.2.3
Sterling exchange rate index	Sterling ERI against EMS currencies, based on 1975 = 100	UK	Monthly average	[QRL.33]	8.2.1.1
Sterling exchange rate index	Effective exchange rate indices for sterling and eight other currencies	UK	Weekly, monthly, annual	[QRL.4]	8.2.2
Sterling exchange rate index	Sterling ERI expressed as an index with 1980 = 100	UK	Monthly, quarterly, annual	[QRL.37]	8.2.5
Sterling exchange rate against various other currencies	Exchange rates of 13 major currencies against sterling	UK	Monthly averages	[QRL.33]	8.2.1.1
Sterling exchange rate against various other currencies	Exchange rates of sterling with 15 major currencies. Spot rates and three month forward margins	UK	Monthly	[QRL.33]	8.2.1.2
Sterling exchange rate against various other currencies	Exchange rates against sterling (and the dollar) of seven leading currencies	UK	Weekly, monthly, annual	[QRL.4]	8.2.2
Sterling exchange rate against various other currencies	Sterling exchange rate against six leading currencies	UK	Monthly, quarterly, annual	[QRL.30]	8.2.3

Type of data	Breakdown	Area	Frequency	QRL publication	Text reference
Sterling-dollar exchange rate	Spot rate and three month forward premium	UK	Weekly, monthly, annual	[QRL.4]	8.2.2
Sterling-dollar exchange rate		UK	Daily, monthly, quarterly, annual	[QRL.37]	8.2.5. Daily figures in summary tables, others in UK entry
Sterling-SDR exchange rate		UK	Monthly	[QRL.33]	8.2.1.3
Sterling-SDR exchange rate		UK	Daily, monthly, quarterly, annual	[QRL.37]	8.2.5. Daily figures in summary tables
Sterling-ECU exchange rate		UK	Monthly	[QRL.33]	8.2.1.3
Sterling-ECU exchange rate		UK	Monthly, quarterly, annual	[QRL.37]	8.2.5
UK Foreign Aid					
Main components	Covers three broad aggregates of aid provision: (i) net aid programme, (ii) net public expenditure on aid, (iii) total net official flows. Data is given on sub-components of these, which facilitates the reconciliation of these differing aggregates	UK	Annual	[QRL.6]	9.2. See also 9.9 especially 9.9.4.8
Bilateral flows – gross expenditure	Bilateral aid divided into: (i) financial aid, (ii) technical cooperation, (iii) administrative costs, with further subdivisions	UK	Annual	[QRL.6]	9.3.1.2 to 9.3.1.5
Bilateral flows – gross expenditure	Divided into project aid and non-project aid, with subdivisions	UK	Annual	[QRL.6]	9.3.2.1 to 9.3.2.3

Bilateral flows – gross expenditure	By terms	UK	Annual	[QRL.6]	9.3.2.4. See also 9.9.4.9
Bilateral flows – gross expenditure	By component and detailed geographical distribution	UK	Annual	[QRL.6]	9.7. See also 9.3.3 and 9.9.3
Commitments	By purpose (project, non-project) and terms	UK	Annual	[QRL.6]	9.3.4
Multilateral aid	UK contributions to six groups of agencies	UK	Annual	[QRL.6]	9.3.5. See also 9.3.1.6 and 9.9.4.4
Project aid (expenditure and commitment)	By economic sector	UK	Annual	[QRL.6]	9.3.6
Technical cooperation expenditure	Information on 13 broad types of technical cooperation expenditure, with further subdivisions	UK	Annual	[QRL.6]	9.4
Technical cooperation – manpower data	Data on numbers of personnel involved in technical cooperation, either as experts or students and trainees	UK	Annual	[QRL.6]	9.8
Debt relief	Information on: (i) debt cancellations, (ii) debt rescheduling, (iii) debt refinancing	UK	Annual	[QRL.6]	9.5
Commonwealth Development Corporation		UK	Annual	[QRL.6], [QRL.22]	9.6
Comparative performance	Data comparing UK aid performance in quantitative (volume) and qualitative (terms, tying etc.) aspects with that of other donors	International including UK	Annual	[QRL.66], [QRL.34], [QRL.26]	9.9

Type of data	Breakdown	Area	Frequency	QRL publication	Text reference
Summary Balance of Payments					
Summary Balance of Payments	Visible, Invisibles and Current Balances	UK and Rest of World	Quarterly	[QRL.62]	1.4. Current account items seasonally adjusted. Current balance adjusted and unadjusted. Preliminary figures for latest quarter. Official reserves (drawings on + additions to – this information also appears in [QRL.7]

Balance of Payments Summary	Visible, Invisibles and Current Balances	UK and Rest of World	Quarterly	[QRL.42]	1.4. Current account items seasonally adjusted. Current balance adjusted and unadjusted. This information appears in a very similar table in *Economic Trends* (March, June, September and December) and *Financial Statistics* [QRL.33] Table 10.1. The *Monthly Digest of Statistics* [QRL.42] differs in that (a) it breaks down Official Financing into the above 3 parts and (b) allocation of SDRs column does not also include Gold Subscription to IMF. There are no quarterly entries in SDRs column nor in Net transactions with Overseas Monetary Authorities
Balance of Payments Summary	Visible, Invisibles and Current Balances	UK and Rest of World	Quarterly	[QRL.30]	1.4. Current account items seasonally adjusted. Current balance adjusted and unadjusted. No entries in capital transfers. No entries in allocation of SDRs and Gold Subscription to IMF since 1981 1st quarter. Preliminary estimates for latest quarter. March, June, September and December editions

Type of data	Breakdown	Area	Frequency	QRL publication	Text reference
UK Balance of Payments Summary	Visible, Invisibles and Current Balances	UK and Rest of World	Quarterly	[QRL.33]	1.4. Current account items seasonally adjusted. Current balance adjusted and unadjusted. This information is also given in *Economic Trends* (March, June, September and December). Allocation of SDRs and Gold Subscription to IMF shows only one entry – in first quarter of 1981
Balance of Payments Summary	Current Balance	UK and Rest of World	Quarterly, annual	[QRL.30]	1.4. Annual and quarterly series cover over 25 years. Quarterly series unadjusted. No entry in EEA loss on forwards since 1968. No entry in allocation of SDRs since 1981 1st quarter. Last entry in Gold Subscription to IMF in 1970 4th quarter. Net transactions with Overseas Monetary Authorities, other Monetary Authorities – last entry 1976

Summary Balance of Payments	Current Account	UK and Rest of World	Annual	[QRL.61]	1.4. Capital transfers – no entry since 1974. EEA loss on forward commitments – no entry since 1968. Gold Subscriptions to IMF – no entry since 1970. Allocation of SDRs – no entry since 1981
Balance of Payments of the UK	Current Account: Visible Trade, exports (f.o.b.), imports (f.o.b.), balance. Invisibles, credits, debits, balance. Invisibles Balances, services, IPD, transfers. Current balance	UK and Rest of the World	Annual	[QRL.61]	1.4. This table continued on next three entries.
Balance of Payments of the UK	Capital and Investment transactions	UK and Rest of World	Annual	[QRL.61]	1.4. Capital transfers – no entry since 1974. Investment and other capital transactions – see also next item.
Balance of Payments in the UK	Capital and Investment transactions	UK and Rest of the World	Annual	[QRL.1]	1.4
Balance of Payments in the UK	Allocation of SDRs. Official Financing: Net transactions with Overseas Monetary Authorities. Foreign Currency Borrowing (Net). Official Reserves, drawing on (+) additions to (-). Total official financing. Balancing item	UK and Rest of the World	Annual	[QRL.1]	1.4. This entry is a continuation from the last three entries. No entry in allocation of SDRs since 1981
The Current Account					
Current Balance, Visible Trade and Invisibles	Current Balance. Visible Trade – Exports (f.o.b.), Imports (f.o.b.). Visible Balance, of which Oil, Non-oil. Invisible Balance (projected)	UK and Rest of the World	Monthly	[QRL.7], [QRL.25]	1.4. Seasonally adjusted. Invisibles are projected and subject to revision as more information becomes available for the most recent months

Type of data	Breakdown	Area	Frequency	QRL publication	Text reference
Balance of Payments: Current Account	Visible Trade – Exports (f.o.b.), Imports (f.o.b.), Balance. Invisibles, credits, debits, balances – total, services, IPD, transfers. Current balance	UK and Rest of the World	Monthly	[QRL.30]	1.4. Seasonally adjusted. Invisibles – only given quarterly except Total Balance which is given quarterly and monthly
Balance of Payments: Current Account	Visible Trade – exports, imports, balance. Invisibles – credits, debits, balance. Net seasonal influences on current account. Current balance (unadjusted)	UK and Rest of the World	Monthly	[QRL.42]	1.4. Seasonally adjusted unless otherwise stated. Invisible Credits/debits, Net seasonal influences and unadjusted current balance are only given quarterly
UK Foreign Trade	Exports. Exports less erratic items. Imports. Imports less erratic items. Volume and unit value indices – exports, imports. Terms of Trade – all goods	UK and Rest of the World	Quarterly	[QRL.45]	1.4. Balance of Payments basis. Erratic items = ships, North Sea Oil production installations, aircraft and precious stones. Terms of Trade = export unit values as a percentage of import unit values. Unit values including terms of trade are not seasonally adjusted
Goods and Services: Value and Volume and Price Indices 1980 = 100	Value and Volume Indices, Terms of Trade	UK and Rest of the World	Annual	[QRL.61]	1.4
Balance of Payments: Current Account	Visible and Invisible Exports and Imports	UK and Rest of the World	Quarterly, annual	[QRL.30]	Annual and quarterly series cover over 25 years. Unadjusted and adjusted

Current Account	Visible and Invisible Exports and Imports	UK and Rest of the World	Quarterly	[QRL.62]	1.4. Seasonally adjusted. Preliminary estimates for latest quarter. This information also appears in [QRL.7]
Current Account: Credits, Debits, Balances	Visible and Invisible Exports and Imports	UK and Rest of the World	Annual	[QRL.61]	1.4 Summary table. This table also appears in [QRL.1]
Trade in Oil	Exports and Imports of Oil	UK and Rest of the World	Monthly	[QRL.25]	1.4 Balance of Payments basis seasonally adjusted. This table also appears in [QRL.42]
Visible Trade: Exports and Imports on a Balance of Payments Basis					
Visible Trade	Exports, Imports. Visible Balance	UK and Rest of the World	Monthly	[QRL.25]	1.4. Balance of Payments basis. Seasonally adjusted
Export and Import unit value and volume index numbers 1980 = 100	Export unit value index numbers. Import unit value index numbers. Terms of Trade = export unit value. Index as a percentage of import unit value index. Export volume index numbers. Import volume index numbers	UK and Rest of the World	Monthly	[QRL.25]	1.4. Unit value indices not seasonally adjusted (inc. Terms of Trade). Volume indices seasonally adjusted. Balance of Payments basis. This information also appear in [QRL.7] and in [QRL.30] and also [QRL.42]
Commodity Analysis of Visible Trade	Exports, Imports, Visible Balance: food, beverages and tobacco. Basic materials. Fuels. Semi-manufactures. Finished manufactures. Total manufactures	UK and Rest of the World	Quarterly	[QRL.25]	1.4. Balance of Payments basis

Type of data	Breakdown	Area	Frequency	QRL publication	Text reference
Trade in Goods other than Oil 1980 = 100	Exports and Imports of Non Oil Goods	UK and Rest of the World	Monthly	[QRL.25]	1.1.4 Balance of Payments basis. Erratics = ships, North Sea installations, aircraft, precious stones, silver. All series seasonally adjusted except Unit Value Indices (inc. Terms of Trade). This information is repeated in [QRL.7]
Visible Trade in the Balance of Payments	Total goods: value of exports, value of imports, visible balance. Oil: value of exports, imports, visible balance. Non-oil: value of exports, imports, visible balance	UK and Rest of the World	Quarterly	[QRL.30], [QRL.43]	1.4. Published in March, June, September and December editions. Seasonally adjusted. Total exports and imports also shown unadjusted. [QRL.43] has total exports and imports, adjusted and unadjusted, quarterly, for over 25 years. Visible Trade Balance (seasonally adjusted) also given quarterly in [QRL.33]
Visible Trade of the United Kingdom 1980 = 100	Value of Exports, Imports, visible balance. Unit Value Index numbers: Exports, Imports, Terms of Trade = Export Unit Value Index as a percentage of Import Unit Value Index. Volume Index Numbers: Exports, Imports	UK and Rest of the World	Annual	[QRL.1]	1.4. Balance of Payments basis. *Economic Trends Annual Supplement* [QRL.31] shows value of exports (f.o.b.) and imports (f.o.b.) with visible balance, annually, for over 25 years

Visible Trade: Volume and Unit Value Indices 1980 = 100	Volume Indices: Exports, imports. Unit Value Indices: Exports, imports, Terms of Trade = Unit Value Index for exports. As a percentage of Unit Value Index for imports	UK and Rest of the World	Annual	[QRL.30]	1.4 Balance of Payments basis
Visible Trade	Total exports. Total imports. Visible Balance	UK and Rest of World	Monthly	[QRL.25]	1.4 Seasonally adjusted. This information also appears in [QRL.7]
Trade on a Balance of Payment basis (adjustment from the Overseas Trade Statistics to Balance of Payment basis)	Exports and Imports on Balance of Payments Basis	UK and Rest of the World	Annual	[QRL.61]	1.4
Trade on a Balance of Payments basis: Geographical analysis	Exports, Imports, Visible Balance: European Community, other Western Europe, North America, other developed countries, Rest of World, total	UK and Rest of the World	Annual	[QRL.61]	1.4
Visible Trade on a Balance of Payments basis: Commodity analysis	Exports and Imports on Balance of Payments Basis	UK and Rest of World	Annual	[QRL.61]	1.4
Balance of Payments: Visible Trade					
Value of Exports/Imports	Commodity Classes based on SITC (Rev 2) section, division or group and UN classification by broad economic categories (BEC)	UK and Rest of the World	Monthly	[QRL.25]	1.4.1. These tables re-appear in [QRL.42]
Export/Import Volume Indices 1980 = 100	Commodity Classes based on SITC (Rev 2) section or division and UN classification by BEC	UK and Rest of the World	Monthly	[QRL.25]	1.4.1. These tables re-appear in [QRL.42]
Export/Import Unit Value Indices 1980 = 100	Commodity Classes based on SITC (Rev 2) section or division and UN classification by BEC	UK and Rest of the World	Monthly	[QRL.25]	1.4.1. These tables re-appear in [QRL.42]

Type of data	Breakdown	Area	Frequency	QRL publication	Text reference
Value of Exports/Imports	Areas of the World	UK and Rest of the World	Monthly	[QRL.25]	1.4.1. These tables appear in [QRL.42] which does not include a separate figure for USA. Areas: Developed Countries – Total, European Community, Rest of Western Europe, North America – Total, USA, Other Developed Countries. Developing Countries – total, oil exporting countries, other developing countries. Centrally planned economies
Visible Trade in the Balance of Payments	Exports and Imports. Visible Balance	UK and Rest of the World	Quarterly	[QRL.30]	1.4.1. Seasonally adjusted. Total exports/imports of goods (f.o.b.) also shown unadjusted. March, June, September and December editions
Balance of Payments Current Account: Invisibles					
Invisibles: Summary	Trade in Invisibles, Credit, Debits, Balance	UK and Rest of the World	Quarterly	[QRL.30]	1.4.2. Seasonally adjusted. Total credits, debits and balance shown adjusted and unadjusted. March, June, September and December editions

International Transactions	Exports and Imports of Goods and Services	UK and Rest of the World	Annual	[QRL.64]	1.4. Summary Table. Summary of Balance of Payments in National Accounts form
Balance of Payments: Capital Account					
Balance of Payments – Investment and Other Capital Transactions	Investment and Capital Transactions by Type	UK and Rest of the World	Quarterly	[QRL.42]	1.5. Not seasonally adjusted
Analysis of Capital Transactions (UK Financial Account and Capital Transfers)	Capital Transactions by Type	UK and Rest of the World	Annual	[QRL.61]	1.5
External Assets and Liabilities by Type and Sector	Assets and Liabilities by Type and Sector	UK and Rest of the World	Annual (Year End)	[QRL.61]	1.5
Official Financing					
Official Financing	Total Net transactions with the IMF and other Overseas Monetary Authorities. Foreign currency borrowing – by HM Government, by public bodies under exchange cover scheme. Drawings on (+) or additions to (-) official reserves	UK and Rest of the World	Quarterly	[QRL.33]	1.5.1. No entry in Net transactions with IMF and other monetary authorities since 1983 first quarter. Official financing abandoned in 1986
Official Financing Liabilities and Official Reserves	IMF. Other foreign currency borrowing – by HM Government, by public bodies under exchange cover scheme. Total official financing liabilities. Official reserves – total, Gold, IMF SDRs, convertible currencies, reserve position in IMF	UK and Rest of the World	Quarterly	[QRL.42]	1.5.1. Official reserves items given in \$ million, all other items in £ million, total official reserves given in \$ million and £ million

Type of data	Breakdown	Area	Frequency	QRL publication	Text reference
Transactions with Countries and Institutions of the European Community					
UK Transactions with Countries and Institutions of the European Community: Summary Account	Transactions by type	UK and European Community	Annual	[QRL.61]	1.6
UK Transactions with Countries and Institutions of the European Community: Summary Account	Capital transactions by type, investments, bank deposits and lending reserves	UK and European Community	Annual	[QRL.61]	1.6. Official reserves: additions to (-) drawings on (+).
Trade by Areas of World					
General Summary of the Value of Trade	Total imports and exports	UK and Rest of the World	Monthly	[QRL.48]	2.2
Summary of the Value of Trade	Imports and exports by areas of the world and countries	UK and Rest of the World	Monthly	[QRL.48]	2.4. Areas: European Community, Rest of Western Europe, North America, Other developed countries, Latin America, Middle East and North Africa, Other developing countries, centrally planned economies, EFTA, OECD, Oil exporters, Commonwealth
Value of Exports to/from European Community	In total and by countries	UK and EC	Monthly	[QRL.43]	2.4. Seasonally adjusted.

Value and Quantity of Imports/Exports	Commodities by SITC (Rev 2) groups, subgroups and items (3, 4 and 5 digit code numbers) in total and for European Community and any countries accounting for significant amounts	UK and Rest of the World	Monthly	[QRL.48]	2.5
Value and Quantity of Imports/Exports	Commodities by SITC (Rev 2) 8 digit code numbers for every trade description in the tariff and overseas trade classification	UK and Rest of the World	Monthly	[QRL.48]	Commodity headings in SITC (Rev 2) subgroup or item order (4 or 5 digit code numbers). It was necessary to introduce new trade descriptions during 1984 to take account of changes to the basic tariff structure. Both the original and replacement 8 digit code numbers are included in tables iv and vii. They are identified by a letter symbol immediately following the value figure
Value of Exports/Imports – Areas of the World	Total, developed countries – total, European Community, Rest of Western Europe, North America, other. Developing countries – total, Oil exporting countries, other. Centrally planned economies	UK and Rest of the World	Monthly	[QRL.43]	2.5. Seasonally adjusted
Value of Exports to/from EFTA and North America	Countries	UK, EFTA and North America	Monthly	[QRL.43]	2.5. Seasonally adjusted.
Value of Exports to/from other Developed Countries, Developing Countries and Soviet Union	Countries. Other Developed – Australia, New Zealand, South Africa, Japan. Developing Countries – Latin America, Middle East and North Africa – total, Israel, Iran, Saudi Arabia, Nigeria, Soviet Union	UK and Rest of the World	Monthly	[QRL.43]	2.5. Seasonally adjusted

Type of data	Breakdown	Area	Frequency	QRL publication	Text reference
Value of Exports (f.o.b.) and Imports (c.i.f.)	Areas of the World	UK and Rest of the World	Monthly	[QRL.7]	2.5 The Table is almost identical to [QRL.42] Table 15.4 but [QRL.7] also has data on a quarterly basis as well as monthly. There is a slight difference in areas covered: [QRL.7] shows North America – total and USA; [QRL.42] only shows North America in total. Seasonally adjusted. Areas are: Developed Countries (total, European Community, Rest of Western Europe, North America – total, USA, other). Developing countries (total, Oil exporting countries, other). Centrally planned economies

Value of Exports (f.o.b.)	Areas of the World and Countries	UK and Rest of the World	Annual	[QRL.1]	2.5. Areas are: European Community, Rest of Western Europe, North America, other Developed Countries, Oil exporting countries, other Developing Countries, centrally planned economies. This table contains similar information to Table 1.B in *Overseas Trade Statistics of the UK* [QRL.49] – annual, but there is some difference in areas and countries covered and the length of the series. The *Annual Abstract of Statistics* [QRL.1] gives longer series and covers fewer countries

Type of data	Breakdown	Area	Frequency	QRL publication	Text reference
Value of Imports (c.i.f.)	Areas of the World and Countries	UK and Rest of the World	Annual	[QRL.1]	2.4. Areas are: European Community, Rest of Western Europe, North America, other Developed Countries, Oil exporting countries, other Developing Countries, centrally planned economies. This Table contains similar information to [QRL.48] *Overseas Trade Statistics of the UK* – annual, but there is some difference in the areas and countries covered and the length of the series. [QRL.1] gives a longer series and covers fewer countries
Value of Exports (f.o.b.) and Imports (c.i.f.)	Areas of the World	UK and Rest of the World	Monthly	[QRL.42]	2.4. Areas are: Developed Countries – total, European Community, Rest of Western Europes, North America, other Developed Countries. Developing Countries – total, Oil exporting countries, other Developing Countries. Centrally planned economies. Seasonally adjusted

Value of Exports (f.o.b.)	Countries	UK and Rest of the World	Monthly	[QRL.42]	2.4
Values of Imports (c.i.f.)	Countries	UK and Rest of the World	Monthly	[QRL.42]	2.4
Value of Exports (f.o.b.) and Imports (c.i.f.)	Industries based on SIC (1980) divisions and classes	UK and Rest of the World	Quarterly	[QRL.15]	2.5
Value of Exports (f.o.b.) and Imports (c.i.f.)	Commodities by SITC (Rev 2) sections and divisions (1 and 2 digit code numbers)	UK and Rest of the World	Annual	[QRL.1]	2.5
Value of Exports (f.o.b.) and Imports (c.i.f.)	Commodity classes based on SITC (Rev 2) section, division or group and UN classification by BEC	UK and Rest of the World	Monthly	[QRL.42]	2.5. Seasonally adjusted
Value of Exports (f.o.b.)	Commodities by SITC (Rev 2) sections and divisions (1 and 2 digit code numbers)	UK and Rest of the World	Monthly	[QRL.42]	2.5
Value of Imports (c.i.f.)	Commodities by SITC (Rev 2) sections and divisions (1 and 2 digit code numbers)	UK and Rest of the World	Monthly	[QRL.42]	2.5
Trade by Commodity					
Value of Commodity Exports	Major commodity group based on UN Broad Economic Categories (BEC) and/or SITC (Rev. 2)	UK and Rest of the World	Monthly	[QRL.43]	1.5. Seasonally adjusted. See Appendix 2

Type of data	Breakdown	Area	Frequency	QRL publication	Text reference
Trade by Area and Commodity					
Visible Exports and Imports	Total exports, total imports, exports less imports and exports over imports % of commodities by SITC (Rev 2) section and division headings, (1 and 2 digit code numbers), by European country, Japan, USA, and by area	UK and Rest of the World	Annual	[QRL.63]	2.5.1. Statistics also given cumulatively for the current year and the corresponding months of the previous year. Statistics on an Overseas Trade Statistics basis. Exports (f.o.b.), Imports (c.i.f.). Areas: EC, Rest of Western Europe, EFTA, total Western Europes, OECD, Oil exporters, Latin America, centrally planned economies, total world
Value of Imports/exports	Commodities by SITC (Rev 2) division and sections (1 and 2 digit code numbers) for areas of the world and countries	UK and Rest of the World	Monthly	[QRL.48]	2.5.1. Areas: EC, rest of Western Europe, North America, other Developed Countries, Latin America, Middle East and North Africa, other Developing Countries, centrally planned economies, EFTA, OECD, Oil exporters, Commonwealth
Direct Exports by Nature of Enterprise					
Large Exporters	By size of firm's exports	UK	Annual 1971-81	[QRL.10] Appendix 10	6.4.2

Direct Exports	By category of enterprise, with respect to nature of overseas affiliation (see 6.4.1.4)	UK	Annual 1970-81	[QRL.10], [QRL.60] Appendix 10	6.4.3. 'Related Concern' export data included for 1970, 1973, 1976, 1979 to 1981
Direct Exports	By category of enterprise (see 6.4.1.4) and main industrial activity	UK	Annual 1971-81 and 1966	[QRL.10], [QRL.60], [QRL.5] Appendix 10	6.4.4. 'Related concern' export data for 1966, 1970, 1973, 1976, 1979 to 1981. Covers 'returns received' only
Direct Exports	By category of enterprise (see 6.4.1.4) and by size of exports	UK	Annual 1970-81 and 1966	[QRL.10], [QRL.60], [QRL.5] Appendix 10	6.4.5. 'Related concern' exports data for 1966, 1970, 1973, 1976, 1979 to 1981. Covers 'returns received' only
Monetary Gold					
Value and Quantity of Imports and Exports of Monetary Gold	By inter-bank and London Gold Market dealings in unwrought, refined gold bullion	UK and Rest of the World	Monthly	[QRL.48]	2.6. From January 1981, all trade in gold, other than that in unwrought refined bullion which is the subject of inter-bank or London Gold Market dealings (monetary gold) is included in Merchandise Trade
Invoicing Currency					
Invoicing Currency of Exports and Imports	Area, commodity and size of transaction	UK and Rest of the World	Biannual	[QRL.7]	2.9
Trade Volume and Unit Value Indices					
Commodity Exports – Volume Indices	Major commodity group based on UN Broad Economic Categories (BEC) and/or SITC (Rev. 2)	UK and Rest of the World	Monthly	[QRL.43]	2.10. Seasonally adjusted
Commodity Exports – Unit Value Indices	Major commodity group based on UN Broad Economic Categories (BEC) and/or SITC (Rev. 2)	UK and Rest of the World	Monthly	[QRL.43]	2.10. Seasonally adjusted

Type of data	Breakdown	Area	Frequency	QRL publication	Text reference
Value of Commodity Imports, Commodity Imports Volume and Unit Value Indices	Major commodity group based on SITC (Rev. 2)	UK and Rest of the World	Monthly	[QRL.43]	2.10. Seasonally adjusted except unit value indices
Value of Imports and Imports Volume and Unit Value Indices Food, Beverages, Tobacco and Basic Materials	Food and live animals: total, meat and meat preparations, dairy products and eggs, cereals and animal feeding stuffs, fruit and vegetables, beverages and tobacco – total, beverages, tobacco. Basic materials: crude materials (total, wood, lumber and cork, pulp and waste paper, textile fibres, metal ores) – animal and vegetable oils and fats	UK and Rest of the World	Monthly	[QRL.43]	2.10. Seasonally adjusted except unit value indices
Value of Commodity Imports and Commodity Imports Volume and Unit Value Indices	Semi-manufactured goods, finished manufactured goods, finished manufactured goods excluding erratics, based on SITC (Rev. 2) and UN Broad Economic Categories (BEC).	UK and Rest of the World	Monthly	[QRL.43]	2.10. Seasonally adjusted except unit value indices
Value of Imports and Exports	Industries by SIC (1980) activity headings	UK and Rest of the World	Quarterly	[QRL.15]	2.10. Imports (c.i.f.) exports (f.o.b.)
Export and Import Volume Index Numbers	Commodity classes based on SITC (Rev 2) section or division and UN classification by BEC	UK and Rest of the World	Monthly	[QRL.42]	2.10. Seasonally adjusted
Export and Import Unit Value Index Numbers	Commodity classes based on SITC (Rev 2) section or division and UN classification by BEC	UK and Rest of the World	Monthly	[QRL.42]	2.10
Volume and Unit Value Indices for Imports	Commodity classes based on SITC (Rev 2) section or division	UK and Rest of the World	Quarterly	[QRL.45]	2.10. Seasonally adjusted except import unit values

Value of Exports and Export Volume Index Numbers	Value by areas of the world volume index by commodity classes based on SITC (Rev 2) section or division	UK and Rest of the World	Quarterly	[QRL.45]	2.10. Areas: developed countries (total, EC, EFTA, North America) and developing countries (total, oil exporting countries). Seasonally adjusted
Import Penetration and Export Sales Ratios					
Import Penetration Ratios	By SIC class and group	UK	Quarterly	[QRL.16]	2.11
Import Penetration and Export Sales Ratios	Products of manufacturing industries by SIC (1980) divisions and classes	UK and Rest of the World	Quarterly	[QRL.7]	2.11. Series covers previous months ending latest quarter.
Import Penetration and Export Sales Ratios	Products of manufacturing industries and classes	UK and Rest of the World	Quarterly	[QRL.42]	2.11. Series covers previous twelve months ending latest quarter
Import Penetration and Export Sales Ratios	Products of Manufacturing Industry by SIC (1980) classes	UK and Rest of the World	Annual	[QRL.1]	2.11, 2.12. An article in *Economic Trends* [QRL.30] August 1977 is concerned with the ratios and problems associated with them. The latest ratios for full detail within manufacturing (over 200 activity headings) are shown in *Business Monitor MQ12, Import Penetration and Export Sales Ratios for Manufacturing Industry* [QRL.16].
Export Performance Ratios	By SIC class and group	UK	Quarterly	[QRL.16]	2.11

Type of data	Breakdown	Area	Frequency	QRL publication	Text reference
Competitiveness					
Measures of UK Competitiveness in Trade in Manufactures	Summary measures: relative export prices, relative producer prices, IMF index of relative unit labour costs – actual, normalised. Import price competitiveness. Relative profitability of exports. Export unit value index, producer price index, unit labour costs index: UK, US, Japan, France, Federal Republic of Germany	UK and Rest of the World	Quarterly	[QRL.30]	2.12
Scotland					
Value of Imports and Exports through Scottish Ports	Commodity classes based on SITC (Rev 2) division and section titles	Scotland and Rest of the World	Annual	[QRL.56]	2.13
Northern Ireland					
Northern Ireland – Inward Investment	Inward investment in Northern Ireland: number of jobs – cost per job, source nation	NI	Annual	[QRL.36], [QRL.41]	2.14
Northern Ireland – Trade	Import/export figures until end 1973 by area and commodity. Series of tonnage of goods imported/exported through principal ports ends 1980	NI	Discontinued	[QRL.46]	2.14. From 1980 onwards, no published data on imports/exports for Northern Ireland
Northern Ireland – Tourism	Passenger travel from Northern Ireland by mode of transport. Home holidays spending and total revenue	NI	Annual	[QRL.46]	2.14

Oil					
Arrivals and Disposals of Norwegian-owned Crude Oil from the Continental Shelf	Arrivals in total, disposals by country	UK and Rest of the World	Monthly	[QRL.48]	2.15. Oil which arrives in the UK and also appears as "Disposals to the UK" is treated as an import and appears in import figures in other tables. Oil which arrives in the UK but is disposed of to other countries is regarded as being in transit
Customs Duties					
Customs Duties: Analysis of Total Imports	Commodities by SITC (Rev 2) section (1 digit code numbers), showing dutiable status and duty charged for imports from the whole world, from the EC	UK and Rest of the World	Annual	[QRL.54]	2.16
Invisible Trade —General					
Balance of Payments Current Account: Invisibles Summary	Credits, Debits and Balances, Services, IPD and Transfers, Total	UK and Rest of the World	Annual	[QRL.61]	3.1
Balance of Payments Current Account: Invisible Credits and Debits	Invisible Credits and Debits by Type	UK and Rest of the World	Quarterly	[QRL.42]	3.1. Similar to [QRL.30] but has seasonally adjusted and unadjusted figures and is quarterly
Services Summary	Credits, Debits and Balances: private sector and public corporations	UK and Rest of the World	Quarterly	[QRL.25]	3.1. Preliminary figures for most recent quarter
Invisibles seasonally adjusted	All sectors: credits, debits, balance – of which services, IPD, transfers. Private sector and public corporations: credits, debits, balance	UK and Rest of the World	Quarterly	[QRL.25]	3.1. Private sector and public corporations, excludes general government transactions and all transfers

Type of data	Breakdown	Area	Frequency	QRL publication	Text reference
Total invisible receipts (excluding miscellaneous government receipts).	By country for top '35' and total in aggregate	World	Annual	[QRL.67]	3.1. See Appendix 4 for top '35'
Total invisible payments (excluding miscellaneous government payments)	By country for top '35' and total in aggregate	World	Annual	[QRL.67]	3.1
Invisible trade as a proportion of total trade and GNP	By country for top '35' and aggregates	World	Annual	[QRL.67]	3.1
Composition of invisible trade – receipts/payments	By main category and by country for top '35' and aggregates	World	Annual	[QRL.67]	3.1. Main categories: transport, travel, investment income, other services, total
National Balances of Invisible Trade. Growth Index of Total Invisibles	By main category and by country for top '35' and aggregates	World	Annual	[QRL.67]	3.1
Services: Volume and Implied Price Indices	Volume Index: Exports/imports of services seasonally adjusted. Implied Price Index: Exports/imports of services. Unadjusted	UK and Rest of the World	Quarterly	[QRL.30]	3.1. Preliminary figures for latest quarter
Invisibles: Summary	Credits/debits: Services, IPD, transfers, total. Invisibles Balance: Total, by sector	UK and Rest of the World	Quarterly	[QRL.30]	3.1. Seasonally adjusted. Total credits, debits and balance also given unadjusted. Preliminary figures for most recent quarter

Sea Transport					
Services: Sea Transport	Ships owned or chartered by UK residents, ships operated by overseas residents. Dry, wet cargo. Freight on UK exports, cross trades, imports. Overseas, UK passenger revenue. Time charter receipts, charter payments. Disbursements in UK, overseas. Total credits, debits, overall balance	UK and Rest of the World	Annual	[QRL.61]	3.2.3
Civil Aviation					
Services: Civil Aviation	By type of airline and passenger/freight	UK and Rest of the World	Annual	[QRL.61]	3.4
Travel and Tourism					
Services: Travel	Credits, Debits, Balances: by purpose of visit, business, leisure, total	UK and Rest of the World	Annual	[QRL.61]	3.2.5
Visits to UK by Overseas Residents, Visits Abroad by UK Residents	By area of residence, area visited. Total. Percentage change in total on previous year	UK and Rest of the World	Quarterly	[QRL.13]	3.2.5. Areas: North America, EC, Other Western Europe, Other Areas. Total adjusted and unadjusted
Overseas Travel and Tourism: Visits to the UK. Visits Abroad	By country of residence. By mode of travel (air, sea) and purpose of visit (holiday, business, VFR, other). By country visited. By mode of travel and purpose of visit	UK and Rest of the World	Quarterly	[QRL.32]	3.2.5 Provisional for most recent quarter. See [QRL.13] which has same information but slightly more detailed analysis of purpose (includes excursionists as separate category and inclusive tour holidays as separate category)

Type of data	Breakdown	Area	Frequency	QRL publication	Text reference
Overseas Travel and Tourism: Visits to the UK by Overseas Residents, Visits Abroad by UK Residents	By area of residence, area visited. Total	UK and Rest of the World	Monthly	[QRL.32]	3.2.5. Areas: all areas, North America, Western Europe, Other. Totals also given seasonally adjusted. Also given annually.
Overseas Travel and Tourism: Earnings and Expenditure	Total earnings from overseas visitors to the UK, expenditure by UK residents abroad and balance. Adjusted and unadjusted	UK and Rest of the World	Monthly	[QRL.32]	3.2.5. Also annual figures
Overseas Travel and Tourism – Overseas Visitors to the UK, Visits Abroad by UK Residents	By area of residence, area visited	UK and Rest of the World	Monthly	[QRL.7]	3.2.5. Initial estimates
Overseas Earnings and Expenditure	Overseas visitors to UK, UK residents going abroad. At current and at constant 1980 prices. Balance (£m) at current prices. Adjusted and unadjusted	UK and Rest of the World	Quarterly	[QRL.13]	3.2.5 Regions: total, North America, EC, Other Western Europe, Other Areas – as before
Overseas Earnings and Expenditure	Overseas visitors to UK, UK residents going abroad, balance by region of world	UK and Rest of the World	Quarterly	[QRL.13]	3.2.5 Regions: total, North America, EC, Other Western Europe, Other Areas – as before
Visitor – Nights in UK and Abroad by Overseas and UK Residents	By area of residence, area visited. Total. Percentage in total on previous year	UK and Rest of the World	Quarterly	[QRL.13]	3.2.5. Areas: North America, EC, Other Western Europe, Other Areas – as before
Number of Overseas Visits to UK	By country of residence	UK and Rest of the World	Quarterly	[QRL.13]	3.2.5

Number of visits to the UK by overseas residents by all modes of transport	By main purpose of visit and area of residence. Total	UK and Rest of the World	Quarterly	[QRL.13]	3.2.5. Areas as before. Purpose: All visits (of which excursionists); Holiday (of which inclusive tour); Business; VFR (visits to friends and relatives); Miscellaneous
Number of visits abroad by UK residents by all modes of travel	By main purpose of visit and area visited. Total	UK and Rest of the World	Quarterly	[QRL.13]	3.2.5. Areas as before. Purpose analysis as before
Number of overseas visitors to the UK by air/sea	By air/sea by main purpose of visit and area of residence	UK and Rest of the World	Quarterly	[QRL.13]	3.2.5. Areas as before. Purpose analysis as before
Number of overseas visitors to the UK	By country of residence and mode of travel (air, sea). Total	UK and Rest of the World	Annual	[QRL.13]	3.2.5
Overseas visitors to the UK – number of overnight visits to the regions of the UK	By area of residence	UK and Rest of the World	Quarterly	[QRL.13]	3.2.5. Areas as before
Number of visits abroad by UK residents by air/sea	By air/sea by main purpose of visit and area visited	UK and Rest of the World	Quarterly	[QRL.13]	3.2.5. Areas as before. Purpose analysis as before
Number of visits abroad by UK residents	By country visited and mode of travel (air, sea). Total	UK and Rest of the World	Annual	[QRL.13]	3.2.5
Overseas travel and tourism – number of overseas visitors to the UK, visits abroad by UK residents, earnings and expenditure	Overseas visitors to the UK, visits abroad by UK residents by area of residence, area visited. Total earnings, expenditure and balance	UK and Rest of the World	Monthly	[QRL.13]	3.2.5. Areas: North America, Western Europe, Other Areas

Type of data	Breakdown	Area	Frequency	QRL publication	Text reference
Overseas travel and tourism – overseas earnings and expenditure at current prices – initial estimates	Total earnings, expenditure and balance. Adjusted and unadjusted	UK and Rest of the World	Monthly	[QRL.7]	3.2.5
Overseas travel and tourism – overseas visitors to the UK, visits abroad by UK residents	By area of residence, area visited. Total	UK and Rest of the World	Quarterly	[QRL.7]	3.2.5. Areas: North America, EC, Other Western Europe, Other Areas. Total (not seasonally adjusted and seasonally adjusted). Provisional figures for most recent quarter.
Overseas visitors to the UK, visits abroad by UK residents	By mode of travel (air, sea) and purpose of visit (holiday, business, VFR, other) total	UK and Rest of the World	Quarterly	[QRL.7]	3.2.5. Provisional figures for most recent quarters. This information also appears in [QRL.13]
Overseas travel and tourism – overseas earnings and expenditure at current prices	Total earnings, expenditure and balance. Adjusted and unadjusted	UK and Rest of the World	Quarterly	[QRL.7]	3.2.5. Provisional figures for most recent months. This information also given in [QRL.13]
Overseas travel and tourism – visitor nights	Total by overseas visitors to the UK, UK residents going abroad	UK and Rest of the World	Quarterly	[QRL.7]	3.2.5 Provisional figures for latest quarters
Overseas travel and tourism. Summary	Number of visits (000's) to the UK, overseas by overseas visitors, UK residents and total earnings, expenditure and balance (£m). Percentage increase on previous year	UK and Rest of the World	Annual	[QRL.11]	3.2.5

Overseas visitors to the UK – number of visitors and expenditure	By country of permanent residence and mode of transport (all routes, air, sea) By quarter, area of permanent residence and main purpose of visit (holiday, business, VFR, miscellaneous)	UK and Rest of the World	Annual	[QRL.11]	3.2.5. Areas: All Areas, North America, EC, Other Western Europe, Other areas
Overseas visitors to the UK – average length of stay (days), expenditure per visit and per day	By country of permanent residence. By area of permanent residence and main purpose of visit	UK and Rest of the World	Annual	[QRL.11]	3.2.5. Areas and purpose as before in [QRL.11]
Overseas visitors to the UK – number of visits	By area of permanent residence, main purpose of visit and mode of travel (all routes, air, sea). By country of residence and main purpose of visit	UK and Rest of the World	Annual	[QRL.11]	3.2.5. Areas as before
Overseas visitors to the UK – number of overnight visits to the regions of the UK	By country of residence. Total, London, Other England, Scotland and Wales	UK and Rest of the World	Annual	[QRL.11]	3.2.5
Overseas visitors to UK – internal expenditure percentage	By type of expenditure and purpose of visit	UK and Rest of the World	Annual	[QRL.11]	3.2.5. Type of expenditure – shopping: total, clothes, other; accomodation; eating out; travel within UK; services; entertainment; other items
Overseas visitors to UK	By region and countries of UK and leading country of residence	UK and Rest of the World	Annual	[QRL.11]	3.2.5
Visits abroad by UK residents – Number of visits and expenditure abroad	By main country visited and mode of travel (all routes, air, sea). By quarter, area visited and main purpose of visit	UK and Rest of the World	Annual	[QRL.11]	3.2.5. Areas and purpose as before in [QRL.11]; i.e. All Areas, North America, EC, Other Western Europe, Other Areas. Purpose: holiday, business, VFR, miscellaneous

Type of data	Breakdown	Area	Frequency	QRL publication	Text reference
Visits abroad by UK residents – average length of stay expenditure per visit and per day	By main country visited. By area visited and main purpose of visit	UK and Rest of the World	Annual	[QRL.11]	3.2.5. Areas and purpose as before in [QRL.11]
Visits abroad by UK residents – number of visits	By area visited, main purpose of visit and mode of travel (all routes, air, sea). By main country visited and main purpose of visit	UK and Rest of the World	Annual	[QRL.11]	3.2.5. Areas as before
Financial Services					
Services: financial and other services	Detailed analysis of earnings by type of institution, firm or service.	UK and Rest of the World	Annual	[QRL.61]	3.2.6
Overseas earnings of UK financial institutions (The City)	By type of institution. Insurance and banking institutions by type of income	UK and Rest of the World	Annual	[QRL.61]	3.4
Other Services					
Overseas transactions in film and television material	Receipts, expenditure and receipts less expenditure for film companies, television companies and in total	UK and Rest of the World	Annual	[QRL.10]	3.2.6.2
Overseas transactions in film and television material: overseas transactions of the BBC and IBA programme contractors	Receipts and expenditure by area (EC, Other Western Europe, North America, Other Developed Countries, Rest of World, Total) and type of transaction: performances in cinemas or on television. Production of films etc	UK and Rest of the World	Annual	[QRL.10]	3.2.6.2

Overseas transactions in film and television material: overseas transactions in cinematograph films	Receipts and expenditure by area (EC, Other Western Europe, North America, Other Developed Countries, Rest of the World, Total) and type of transaction: performances in cinemas or on television. Production of films, etc	UK and Rest of the World	Annual	[QRL.10]	3.2.6.2
Overseas transactions in films and television material: overseas transaction in cinematograph films	Receipts and expenditure by type of transaction (performances in cinema or on television, production of films etc) and type of company (UK subsidiaries of the major UK film companies, other total)	UK and Rest of the World	Annual	[QRL.10]	3.2.6.2
Royalties					
Overseas royalties and similar transactions (excluding oil companies, banks and insurance companies)	Transactions by type	UK and Rest of the World	Annual	[QRL.10]	3.2.6.2
Overseas royalties and similar transactions (excluding oil companies, banks and insurance companies): receipts and expenditure (summary of returns received)	By area, country and type of royalty	UK and Rest of the World	Annual	[QRL.10]	3.2.6.2. Areas: Western Europe – Total, EC, EFTA, Other Western Europe, North America, Other Developed Countries. Rest of the World – Total, Africa, Asia – Total, Middle East, Other Asia, Caribbean, Central and South America, Other. World – Commonwealth, Developing Countries, Oil Exporting Countries

Type of data	Breakdown	Area	Frequency	QRL publication	Text reference
Overseas royalties and similar transactions (excluding oil, banks and insurance companies): receipts and expenditure (summary of returns received)	By main activity of UK business: receipts; expenditure: total, related, unrelated by SIC industry group	UK and Rest of the World	Annual	[QRL.10]	3.2.6.2
Earnings on External Assets and Liabilities					
Summary of earnings on UK external assets and liabilities	Earnings on direct and portfolio investment and other earnings by type of asset, by sector	UK and Rest of the World	Annual	[QRL.61]	3.3. Total earnings on assets of UK private sector distinguishes between banks and non-bank private sector
Interest, profits and dividends: credits	By type of credit	UK and Rest of the World	Quarterly	[QRL.30]	3.3. Similar to [QRL.62] but includes unadjusted figures, oil/non-oil and official reserves
Interest, profits and dividends: debits and balance	By type of debit	UK and Rest of the World	Quarterly	[QRL.30]	3.3. Excluding loans to HMG and under the public sector exchange scheme which form part of official financing
Interest, profits and dividends	Credits by type	UK and Rest of the World	Quarterly	[QRL.62]	3.3. Preliminary figures for most recent quarter. Seasonally adjusted

Interest profits and dividends	Debits by type	UK and Rest of the World	Quarterly	[QRL.62]	3.3. Excluding loans to HMG and under the public sector exchange scheme which are included under other liabilities. Preliminary figures for most recent quarter. Seasonally adjusted
Interest, profits and dividends	Total balance of which: private sector and public corporations, general government	UK and Rest of the World	Quarterly	[QRL.62]	3.3. Preliminary figure for most quarter.
Earnings on portfolio investment	By type of security	UK and Rest of the World	Annual	[QRL.61]	3.3
Earnings on Direct Investment					
Earnings on assets and liabilities: earnings on direct investment	Earnings on assets and liabilities by type and recipient	UK and Rest of the World	Annual	[QRL.61]	3.3.1
Earnings on direct investment: sector analysis	By sector.	UK and Rest of the World	Annual	[QRL.61]	3.3.1
Outward direct investment overseas by UK companies	By components including net profits	UK and Rest of the World	Annual	[QRL.10]	3.3.1. Excluding oil companies
Outward direct investment overseas by UK companies	By type of receipt/payment and by area	UK and Rest of the World	Annual	[QRL.10]	3.3.1. Excluding oil companies
Inward direct investment in the UK in all industries, other than oil companies	By companies	UK and Rest of the World	Annual	[QRL.10]	3.3.1
Transfers					
Transfers	By type	UK and Rest of the World	Annual	[QRL.61]	3.5

Type of data	Breakdown	Area	Frequency	QRL publication	Text reference
Transfers	Credits and Debits by Type	UK and Rest of the World	Quarterly	[QRL.30]	3.5. This is very similar to [QRL.62] – see below – but includes unadjusted figures and a more detailed analysis of general government receipts from EC institutions
Transfers	Credits and Debits by Type	UK and Rest of the World	Quarterly	[QRL.62]	3.5. Figures from fourth quarter of 1985 onwards are seasonally adjusted
General government transfers: subscriptions and contributions to international organisations	Payments to the EC etc in total by Type	UK and the EC	Annual	[QRL.61]	3.5. Data may be imcomplete: are showing payments into and out of the CF and NLF, but other payments may be relevant
Services: General government	By Type	UK and Rest of the World	Annual	[QRL.61]	3.5
General government transactions with the institutions of the European Community: Services and transfers	By Type	UK and Rest of the World	Annual	[QRL.61]	3.5

Payments to European Communities etc	Budget of the EC. European Investment Bank: capital subscription. Total payments to European Communities etc	UK and EC	Annual	[QRL.24]	3.5.2. As well as above payments which are made from CF Standing Services *additional* payments are made from CF Supply Services. Specific details are given beneath the figures of the Budget of the EC breakdown. These additional payments are *not* part of "Payments to the EC etc"

Type of data	Breakdown	Area	Frequency	QRL publication	Text reference
Payments to the European Communites etc: Budget of the EC	Payments by Type	UK and EC	Annual	[QRL.24]	3.5.2. Item overdrafts under article 12(2) does not appear in previous year (1983-4), otherwise categories identical in previous year. See also Table 10.2 of *UK Balance of Payments* [QRL.61] which deals with EC Budget. The above payments are made from CF Standing Services. Additional payments are made from CF Supply Services but are *not* included in "Payments to the EC etc". CF and NLF accounts for year ended 31st March 1985
Investment and Other Capital Transactions					
Investment and other capital transactions	By type of transaction	UK and Rest of the World	Quarterly	[QRL.33]	4.2.1
International Migration					
International migration	UK Net migration – time series from International Passenger Survey (IPS), Home Office Statistics	UK and World	Annual	[QRL.38]	10.1, 10.2
International migration	UK and World	UK and World	Quarterly	[QRL.51], [QRL.47]	10.3

QUICK REFERENCE LIST KEY TO PUBLICATIONS

Reference	Author or Organisation	Title	Publisher	Frequency or Date	Remarks
[QRL.1]	Central Statistical Office	*Annual Abstract of Statistics*	HMSO, London	Annual	
[QRL.2]	British Invisible Exports Council	*Annual Report*	British Invisible Exports Council	Annual	1984-5 Enquiries to British Invisible Exports Council, Tel: 071-929-0918
[QRL.3]	International Monetary Fund	*Balance of Payments Statistics Yearbook* (parts 1 and 2)	International Monetary Fund	Annual	
[QRL.4]	Bank of England	*Bank of England Quarterly Bulletin*	Bank of England	Quarterly	
[QRL.5]	Board of Trade	*Board of Trade Journal*	HMSO London	Weekly	Last issue 14th October 1970; incorporated into *Trade and Industry* [QRL.33]
[QRL.6]	Overseas Development Administration	*British Aid Statistics*	Government Statistical Service	Annual	
[QRL.7]	Department of Trade and Industry	*British Business*	Department of Trade and Industry	Weekly	Available from British Business subscription department, Oakfield House, Perrymount Road, Haywards Heath, West Sussex RH16 3DH (July 1986). Originally overseas edition of *Trade and Industry* [QRL.33] which it replaced in January 1980
[QRL.8]	Overseas Development Administration	*British Overseas Aid*	Overseas Development Administration	Annual	

Reference	Author or Organisation	Title	Publisher	Frequency or Date	Remarks
[QRL.9]	British Tourist Authority	*The British Home Tourism Survey*	British Tourist Authority	Annual	A continuous sample survey of tourists, covers all Tourist Boards. The survey (and more detailed results available on written request) from British Tourist Authority, Thames Tower, Black's Road, London W6 9EL
[QRL.10]	Business Statistics Office	*Business Monitor MA4, Overseas Transactions*	HMSO, London	Annual	Information and individual copies from the Library, Business Statistics Office, Government Buildings, Cardiff Road, Newport, Gwent NPT 1XG
[QRL.11]	Business Statistics Office	*Business Monitor MA6, Overseas Travel and Tourism*	HMSO, London	Annual	Information and individual copies from the Library, Business Statistics Office, Government Buildings, Cardiff Road, Newport, Gwent NPT 1XG. Gives annual figures for overseas travel and tourism, has more detailed analysis of expenditure than MQ6
[QRL.12]	Department of Transport	*Business Monitor MA8, Nationality of Vessels in United Kingdom Seaborne Trade*	HMSO London	Annual	

[QRL.13]	Business Statistics Office	*Business Monitor MQ6, Overseas Travel and Tourism*	HMSO, London	Quarterly	Information and individual copies from the Library, Business Statistics Office, Government Buildings, Cardiff Road, Newport, Gwent NPT 1XG
[QRL.14]	Business Statistics Office	*Business Monitor MQ7*	HMSO, London	Quarterly	See [QRL.5]
[QRL.15]	Business Statistics Office	*Business Monitor MQ10*	HMSO, London	Quarterly	Individual Monitors from The Librarian, BSO, Government Buildings, Cardiff Road, Newport, Gwent NP9 1XG; Subscription enquiries to HMSO, PO Box 276, London SW8 5DT
[QRL.16]	Department of Trade and Industry	*Business Monitor MQ12 Import Penetration and Export Sales Ratios for Manufacturing Industry*	HMSO London	Quarterly	
[QRL.17]	Department of Trade and Industry –Business Statistics Office	*Business Monitor PA1002 – Report on the Census of Production –Summary Volume*	HMSO London	Annual	
[QRL.18]	Department of Trade and Industry –Business Statistics Office	*Business Monitor PA1003 – Analyses of United Kingdom Manufacturing (Local) Units by Employment Size*	HMSO London	Annual	
[QRL.19]	Civil Aviation Authority	*Civil Aviation Authority Annual Statistics*	Civil Aviation Authority	Annual	
[QRL.20]	Civil Aviation Authority	*Civil Aviation Authority Monthly Statistics*	Civil Aviation Authority	Monthly	
[QRL.21]	Central Statistical Office	*Commodity Flow Accounts for the United Kingdom*	CSO	Annual	Privately financed. CSO, Cabinet Office, Great George Street, London SW1P 3AQ

Reference	Author or Organisation	Title	Publisher	Frequency or Date	Remarks
[QRL.22]	Commonwealth Development Corporation	*Commonwealth Development Corporation –Annual Report and Statement of Accounts*	Commonwealth Development Corporation	Annual	
[QRL.23]	HM Treasury	*Consolidated Fund and National Loans Fund Accounts* House of Commons Paper	HMSO, London	Annual	HMSO and other bookshops (No 116 for year ending March 31st 1985)
[QRL.24]	HM Treasury	*Consolidated Fund and National Loans Fund Accounts Supplementary Statements* House of Commons Paper	HMSO London	Annual	HCP No 117 for year ended 31st March 1985
[QRL.25]	Department of Trade and Industry, Central Statistical Office	DTI Press Notice –*The Current Account of the UK Balance of Payments*	Department of Trade and Industry, London	Monthly	3.1 Exports (FOB) Imports (CIF); seasonally adjusted. For list of commodity classes used see sheet attached to Monthly Digest.
[QRL.26]	Development Assistance Committee of the OECD	*Development Cooperation*	OECD	Annual	Title 'Development Assistance' until 1971 edition
[QRL.27]	British Tourist Authority	*Digest of Tourist Statistics*	British Tourist Authority	Periodically	Most recent edition No 12 (June 1988). British Tourist Authority, Research services, Thames Tower, Black's Road, London W6 9EL
[QRL.28]	The Treasury	*Economic Progress Report*	The Treasury	Monthly	
[QRL.29]	The Working Group of the National Tourist Organisations of the EEC	*The Economic Significance of Tourism within the European Community* A Report	British Tourist Authority, London	Fifth Report 1983	
[QRL.30]	CSO	*Economic Trends*	HMSO, London	Monthly	

[QRL.31]	CSO	*Economic Trends Annual Supplement*	HMSO, London	Annual	
[QRL.32]	Department of Employment	*Employment Gazette*	HMSO, London	Monthly	
[QRL.33]	CSO	*Financial Statistics*	HMSO, London	Monthly	
[QRL.34]	OECD	*Geographical Distribution of Financial Flows to Developing Countries*	OECD	Annual	
[QRL.35]	International Chamber of Commerce (ICC) UK	*Incoterms*	ICC, UK	Revised regularly	Available form ICC, UK. Centre Point, 102 New Oxford St, London WC1A 1QB
[QRL.36]	Northern Ireland Industrial Development Board (IDB)	*IDB Annual Report*	IDB	Annual	Deals with firms in Northern Ireland with more than fifty employees
[QRL.37]	International Monetary Fund	*International Financial Statistics*	International Monetary Fund	Monthly	
[QRL.38]	Office of Population Censuses and Surveys	*International Migration* Series MN	HMSO, London	Annual	Series MN – 1985 is Number 12, includes International Passenger Survey (IPS)
[QRL.39]	Dept. of Transport	*International Road Haulage Survey*	HMSO, London	Annual	Covers UK HGV Vehicles only which cross by ferry (excludes Irish land boundary).
[QRL.40]	Dunning, J. and Cantwell, J.	*IRM Directory of Statistics and International Investment and Production.*	Macmillan Reference Books, London.		
[QRL.41]	Northern Ireland Local Enterprise Development Unit (LEDU)	*LEDU Annual Report*	LEDU	Annual	Deals with firms in Northern Ireland with less than fifty employees
[QRL.42]	Central Statistical Office	*Monthly Digest of Statistics*	HMSO, London	Monthly	FOB and CIF where specified (seasonally adjusted where specified).

Reference	Author or Organisation	Title	Publisher	Frequency or Date	Remarks
[QRL.43]	Department of Trade and Industry	*Monthly Review of External Trade Statistics*	Department of Trade and Industry, London	Monthly	
[QRL.44]	Department of Trade and Industry	*Monthly Review of External Trade Statistics* Annual Supplement	Department of Trade and Industry, London	Annual	Provides longer historic runs of data, as available, for all series covered in the Monthly Edition. Available from Department of Trade and Industry, S2A1, Room 255, 1 Victoria Street, London SW1H 0ET; tel 071-215-4895 /4894
[QRL.45]	National Institute of Economic and Social Research	*National Institute Economic Review*	National Institute of Economic and Social Research	Quarterly	
[QRL.46]	Policy Planning and Research Unit, Department of Finance and Personnel, Stormont	*Northern Ireland Annual Abstract of Statistics*	Policy Planning and Research Unit, Department of Finance and Personnel, Stormont	Annual	
[QRL.47]	Office of Population Censuses and Surveys (OPCS)	*OPCS Monitors*	OPCS		Reference MN, Information Branch (Dept M), OPCS, St Catherine's House, 10 Kingsway, London WC2B 6JP
[QRL.48]	Department of Trade and Industry	*Overseas Trade Statistics of the United Kingdom*	HMSO, London	Monthly	
[QRL.49]	Department of Trade and Industry	*Overseas Trade Statistics of the United Kingdom*	HMSO, London	Annual	

[QRL.50]	Department of Employment	Press Notice *Overseas Travel and Tourism*	Department of Employment	Monthly	This Press Notice formerly came from the Department of Industry
[QRL.51]	Office of Population Censuses and Surveys	*Population Trends*	HMSO	Quarterly	
[QRL.52]	DTI — Business Statistics Office	*Report on the Census of Production*	HMSO	Annual	A list of Business Monitors make up the full volume. For details see [QRL.53]
[QRL.53]	DTI — Business Statistics Office	*Report on the Census of Production Summary Volume*	HMSO	Annual	Business Monitor PA 1002 [single volume]
[QRL.54]	HM Customs and Excise	*Report of the Commissioners of Her Majesty's Customs and Excise*	HMSO, London	Annual	Command Paper (Cmnd 9655 for Edition ended March 31st 1985)
[QRL.55]	The Scottish Council: Development and Industry	*Scotland's Manufactured Exports*	The Scottish Council: Development and Industry	Annual	Annual since 1984, before that triennial. Available from the Scottish Council: Development and Industry, 23 Chester Street, Edinburgh EH3 7ET
[QRL.56]	Scottish Office	*Scottish Economic Bulletin*	HMSO, Edinburgh	Normally Biannual	Available HMSO Bookshops and other booksellers
[QRL.57]	Special Chemical Statistics Ltd	*Special Chemical Return*	Special Chemical Statistics Ltd, Millbank Tower, Room 1231, Millbank, London SW1P 4QX	Monthly	Access by fee only
[QRL.58]	General Council of British Shipping	*Statistical Brief*	General Council of British Shipping	Quarterly	Enquiries to Statistics Department, General Council of British Shipping, 30-32 St Mary Axe, London EC3A 8ET

Reference	Author or Organisation	Title	Publisher	Frequency or Date	Remarks
[QRL.59]	English Tourist Board Research Services	*Tourism Fact Sheet*	British Tourist Authority	Annual	One 'fact sheet' produced for each tourist region of England
[QRL.60]	Department of Trade and Industry	*Trade and Industry*	HMSO, London	Weekly	First issue 21st October 1970; last issue January 1980 (replaced by *British Business*) [QRL.3]
[QRL.61]	CSO	*United Kingdom Balance of Payments* (CSO Pink Book)	HMSO, London	Annual	
[QRL.62]	Central Statistical Office	CSO Press Notice *UK Balance of Payments*	Central Statistical Office	Quarterly	Enquiries to Central Statistical Office, Gt George Street, London SW1P 3AQ
[QRL.63]	Department of Trade and Industry	*UK External Trade Area x Commodity Analysis*	Department of Trade and Industry	Quarterly	Enquiries to Department of Trade and Industry, 52A1, Room 255, 1 Victoria Street, London SW1H 0ET
[QRL.64]	Central Statistical Office	*United Kingdom National Accounts* ('Blue Book')	HMSO, London	Annual	
[QRL.65]	Department of Transport	*United Kingdom Shipping Industry: International Revenue and Expenditure*	Department of Transport, Publications Sales Unit Building, 1 Victoria Road, South Ruislip, Middlesex MA4 0NX		

[QRL.66]	The World Bank	*World Development Report*	Oxford University Press for the World Bank	Annual	
[QRL.67]	British Invisible Exports Council	*World Invisible Trade*	British Invisible Exports Council	Annual	E7 (July 1986 Edition) British Invisible Exports Council, 6th Floor, Dunster House, 37 Mincing Lane, London EC3R 7BQ

BIBLIOGRAPHY

(Arranged alphabetically by author for named authors and then alphabetically by title)

[B.1] Armstrong, A. (Ed.) *Structural Change in the British Economy 1948-1968* Dept. of Applied Economics, University of Cambridge. Chapman and Hall, London, 1974.

[B.2] Artus, J. R. and McGuirk, A. K. "A Revised Version of the Multilateral Exchange Rate Model" in *IMF Staff Papers* Vol XXVIII, No. 2, June 1981. pp. 275-309.

[B.3] Astin, J. A. and Brady, N. J. "Commodity Flow Accounts for the United Kingdom" in *Economic Trends* 1985. pp. 82-96.

[B.4] Bond, I. D. and Brown, R. N. "Recent Developments in the Terms of Trade" in *Bank of England Quarterly Bulletin* Vol 20, No. 3, September 1980. pp. 295-303.

[B.5] Bosworth, D. L. *Intellectual Property Rights* ESRC/RSS, Reviews of UK Statistical Sources, Vol XIX, Review 32. Pergamon Press, Oxford, 1986.

[B.6] Connor, W. H. P. *The Airport Industry* Brown, Son, Ferguson, Glasgow, 1986.

[B.7] Dogget, E.A. and Cresswell, J.C. "Aspects of United Kingdom Trade Competitiveness" in *Economic Trends* No. 304, February 1979. pp. 79-84.

[B.8] Elis-Williams, D. "Seasonal Adjustment of the Overseas Trade Figures" in *Economic Trends* No. 328, February 1987. pp. 99-103.

[B.9] Finlinson, Colin "Statistics in Customs and Excise" in *Statistical News* No. 76, February 1987. pp. 5-9.

[B.10] Hewer, Alan "Manufacturing Industry in the 70's: An Assessment of Import Penetration and Export Performance" in *Economic Trends* No. 320, June 1980. pp. 97-109.

[B.11] Hughes, Kirsty *Exports and Technology*. Cambridge University Press, 1986.

[B.12] Jones, Graham "How UK's Overseas Trade is Measured" in *British Business*, 8 May 1981.

[B.13] Liston, David and Reeves, Nigel *Invisible Exports and the Service Industries*. Pitman, 1988.

[B.14] Mansell, Ken "UK Visible Trade in the Post-War Years" in *Economic Trends* No. 324, October 1980. pp. 136-150.

[B.15] Mark, J. and Strange, R. *The Food Industries* ESRC/RSS Reviews of UK Statistical Sources. Chapman and Hall, London, (forthcoming)

[B.16] Mort, D. *Sea Transport* ESRC/RSS Reviews of UK Statistical Sources, Vol XIV, Review 25. Pergamon Press, Oxford, 1981.

[B.17] Nawaz, Shuja "Why the Current Account Does Not Balance" in *Finance and Development* (IMF, World Bank) September 1987. pp. 43-45.

[B.18] Panic, M. and Joyce, P. L. "UK Manufacturing Industry: International Integration and Trade Performance" in *Bank of England Quarterly Bulletin* Vol 20, No. 1, March 1980.

[B.19] Pentecost, Eric "The Capital Account of the Balance of Payments" in *Economic Review* Vol 4 No. 3, January 1987. pp. 33-37.

[B.20] Philips, Celia *Civil Aviation* ESRC/RSS Reviews of UK Statistical Sources, Vol X, Review 18. Pergamon Press, Oxford, 1979.

[B.21] Pritchard, Alwyn "Exports and Imports of Services Analysed by Industry" in *Economic Trends* No. 328, February 1981.

[B.22] Rhomberg, R. R. "Indices of Effective Exchange Rates" in *IMF Staff Papers* Vol XXIII, No. 1, March 1976. pp. 88-112.
[B.23] Walker, Graeme "United Kingdom Air Transport: An International Industry" in *Economic Trends* No. 323, September 1980. pp. 95-101.
[B.24] Walker, J. "OECD Benchmark Definition of Foreign Direct Investment" in *Statistical News* No. 61, May 1983. pp. 61.1-61.4.
[B.25] Wells, J.D. and Imber, J.C. "The Home and Export Performance of United Kingdom Industries" in *Economic Trends* No. 286, August 1977. pp. 78-79.
[B.26] International Monetary Fund *Balance of Payments Manual* Fourth Edition. Washington, DC. 1977.
[B.27] "Invisible Earnings of the UK Commodity Markets" in *Bank of England Quarterly Bulletin* March 1977.
[B.28] *Bill of Entry Service* HM Customs and Excise, Southend.
[B.29] Department of Trade "Invisible Earnings: How Much From Japan" in *British Business* 17 April 1981.
[B.30] "Sampling Error of International Passenger Survey" in *British Business* 17 August 1984.
[B.31] Department of Trade and Industry /General Council of British Shipping "UK Shipping: a Geographical Analysis" in *British Business* 13 July 1984.
[B.32] Department of Transport "General Trends in Shipping" in *Business Monitor* MR15. HMSO, London, irregularly (intervals of less than one year).
[B.33] Commission of the European Communities *Com(85)310 Final Completing the Internal Market: White Paper from the Commission to the European Council* Brussels. June 1985. Sold by HMSO.
[B.34] OECD *Detailed Benchmark Definition of Foreign Direct Investment* Paris. 1983.
[B.35] Central Statistical Office "The Monthly Invisible Balance" in *Economic Trends* No. 398, December 1986. p. 87.
[B.36] Department of Trade "Revised Measures of the UK Trade Competitiveness in Manufactures" in *Economic Trends* No. 319, May 1980.
[B.37] International Monetary Fund *Final Report of the Working Party on the Statistical Discrepancy in World Current Account Balance* IMF, Washington, DC. 1987.
[B.38] HM Customs and Excise *Guide to the Classification for Overseas Trade Statistics* HMSO, London. Annual.
[B.39] HM Customs and Excise *HM Customs and Excise Tariff and Overseas Trade Classification* (Tariff). HMSO, London. Annual.
[B.40] OPCS/Home Office *Immigration Statistics: Sources and Definitions* OPCS Occasional Paper 15. 1979. HMSO, London.
[B.41] British Invisible Exports Council *International Information Flows: a Report by the LOTIS Committee* (Liberalisation of Trade in Services), 1986.
[B.42] OPCS *International Migration* 1982. Series MN No. 9, OPCS. HMSO.
[B.43] *United Kingdom National Accounts: Sources and Methods* HMSO, London, 1985.
[B.44] Department of Trade and Industry *United Kingdom National Study of Trade in Services* Department of Trade and Industry. 1984.
[B.45] UN *UN System of National Accounts* Studies in Methods, Series F, No. 2. UN, New York, Irregularly.
[B.46] International Monetary Fund *World Economic Outlook* IMF, Washington, DC. 1987.

ADDENDUM

UK international trade statistics have recently (January 1989) become the subject of controversy. The move to a balance of payments deficit after several years of surplus has focused attention on revisions and the size and possible biases in the balancing item in the accounts. The Statistics User's Council Annual Conference (December 1988) focused attention sharply on the deficiencies in trade statistics against the background of further cuts in the Government Statistical Service. The papers from this Conference are available from the Statistics Users Council Secretary, Mr Ian Maclean, IMAC Research, Lancaster House, More Lane, Esher, Surrey.

This addendum reviews current issues and provides cross references to the text for a discussion of the background and technical details.

1. Revisions to the balance of payments statistics

Attention has focused on revisions to the balance on current account as this has moved to deficit following several years of substantial surplus. The revision, frequently changing deficit into surplus, arises because invisibles are measured after the event through periodical enquiries and revisions occur several years after the reference period (text reference: section 1.4). (See also Bruce Buckingham's paper in the Statistics User's Council 1988 Conference).

2. The balancing item

The persistent positive balancing item in the period 1983 to 1987 has led to the view that there has been a systematic bias in the accounts understating net credits. (text reference: section 1.2.1). The major gaps and imperfections in the UK accounts lay in the field of capital transactions (text reference: section Chapters 4 to 7) particularly in the area of new financial instruments—see below. Capital outflows recorded through the banks and other financial institutions are better recorded than those outside the banking system and the CSO believe that missing inflows are a major cause of the persistent positive balancing item. This view is based on circumstantial evidence however and illustrates the general need for more careful monitoring of key items.

3. The world current account discrepancy

The "world current account" should sum to zero. In practice a persistent and growing deficit has been identified in the sum of current accounts of all countries (text reference: Introduction). This underecording or missing of net credits is likely to be differential across countries and is not guaranteed to adjust the UK current account upwards (if rectified). The impact of suggested IMF revisions is unlikely to have a major impact on UK practices or outcomes.

4. The recording of invisibles

The most contentious and difficult current problem is the recording of invisible transactions in the balance of payments (text reference: section Chapter 3 *passim*).

The quality of information is variable and in certain cases it is absent. The need to rely on intermediaries for information rather than direct collection is an ongoing problem as the information is costly, can be unreliable in coverage and because the intermediary lacks complete knowledge. Changes in asset or liability levels are used to infer transactions in some cases and with new financial instruments being developed, definitions and distinctions become irrelevant. Alan McIntyre of the CSO gives the following breakdown of sources used to compile estimates of services data: Government sources (5% of total transactions), counterparts to visible imports adjustments (7%), specific quarterly surveys (58%), specific annual enquiries (18%), direct calculations (3%), projections based on recent or reasonable figures (5%), projections from more distant base lines (4%). (see UK International Trade Statistics: Statistics User's Council Annual Conference, December 1988). Indications of quality have been removed from the Pink Book [QRL.61] since 1976 [text reference: section 1.3.4] but between 1975 and 1985 the revisions have only exceeded £1 billion once (in 1982). This means that users have to wait until revisions are completed (often a matter of years) before an accurate figure is reached. There is currently no prospect of moving to central recording of all transactions nor of a comprehensive and detailed survey (McIntyre) so piecemeal improvements are being made, largely under pressure from individual users and where known weaknesses are identified.

5. Introduction of the world-wide Harmonised System for the classification of goods

On 1st January 1988 the Harmonised System (HS) classification of goods and the European Community's related integrated tariff (TARIC) were introduced (Text reference: section 2.3.2).

In terms of statistical provision, the change-over was relatively smooth and rebasing of the volume series back to 1985 has been carried out. The introduction of a finer subdivision results in an increase from 9,300 to 15,000 codes based on a framework of eleven digit headings: digits 1-6 HS codes for international use, 7-8 EC codes for CCT and NIMEXE, 9 for use by individual member states to give greater detail for trade or statistical purposes and 10-11 additional TARIC subdivision for tariff measures applicable to EC imports from non-EC countries (David G Mayes and Ulric M Spencer, NEDO in Statistics Users Council 1988 Conference Papers).

6. The Single Administrative Document

The Single Administrative Document introduced on 1st January 1988 in the UK and all other EC countries replaces most of the forms used for customs freight declarations from which all visible trade statistics are derived (text reference: section 2.16). This voluminous form provides information for a variety of purposes and there have been severe teething problems which will gradually work through the statistical system.

Future developments towards the Single European Market of 1992 are likely to create problems for collection of trade statistics, on the assumption that border controls will be abolished. It is likely that surveys of traders plus a possible VAT clearing house system will form the basis of the future system.

7. Deregulation

One marked side-effect of governmental deregulation has been the loss of statistical information. Exchange control records are an obvious case and, as a pillar of information on international transactions, they cannot be adequately replaced by voluntary or even compulsory surveys. Attempts to remove the burden of statistical form filling by businessmen, entrepreneurs, travellers and others inevitably result in an erosion of the quantity and quality of information. Many users believe that this trade-off has swung too far against the needs of statistical coverage and accuracy. Cases of suppression when specific companies could be identified may have been taken too far. The desirability of further disclosure must be constantly kept under review with the users' needs in mind.

The replacement of the bill of entry service in 1986 by marketing agents is an example of privatisation of services and it appears to have led to an increase in the quality of information and range of services. Further developments along the lines of the Special Chemical Return (text reference: section 2.5.3) are to be encouraged and welcomed. The close relationship with users is particularly to be commended.

8. Use of information technology

The increasing use of information technology by providers and users of statistics is increasing the potential for the use of information. Computerisation of Single Administrative Document returns should enable detailed analyses to be developed. It opens the possibility of paperless trading. It is essential that resources are devoted to computerising information and making the information and analyses compatible between user and collector.

9. Miscellaneous problems

The absence of comprehensive geographical analyses of invisible trade data is a frequent complaint of users. Buckingham (1988) *op cit* indicates that the issue is less urgent than ensuring that aggregate figures are securely based. Such figures exist for travel (text reference: section 3.2.5.) royalties and related services, (Chapter 6) film and television services (sections 3.2.6.2 and 6.3) direct investment earnings (sections 3.3.1 and 5.3.5 to 5.3.8). Government services and transfers (sections 3.3.2 and 3.5.1) and direct investment inflows and outflows (sections 4.2.1 and Chapter 5). However the absence of such data negates the possibility of classic market analysis (product x national market). A prerequisite of such an analysis is the achievement of international standards on the classification of services.

Harmonisation of balance of payments data with other statistical information eg. national accounts is also important. Buckingham (1988) points out that balance of payments estimates have historically been concerned with classifying payments and receipts across the exchanges, which is often at conflict with other accounting systems. Work is continuing at an international level to revise the existing frameworks for national accounts (the UN System of National Accounts) and for balance of payments statistics (the IMF Balance of Payments Manual) to make them more closely related. Again, at the centre of the problems is the conceptualisation and measurement of financial services. This may lead to changes in official publications over the next few years.

Finally, a perennial problem is the time taken to resolve trade statistics queries. One hope for resolving this problem is the development of interactive computer systems.

Conclusion

Balance of payments statistics have traditionally relied upon comprehensive data from Customs and Excise and foreign exchange control systems. Changes in the nature and type of trade data arising from harmonisation and reclassification have been dealt with above. The abandonment of formal exchange control does not necessarily mean that exchange records of payments cannot be used as many countries still do. However, the UK attempt to monitor invisibles through the banking system (analagous to the monitoring of visible trade through customs) means a heavy reliance on survey information which is often provided retrospectively.

Whilst the cost of collecting detailed trade statistics must be balanced against their usefulness, it is dismaying to read (*Times* 9.12.88) that the annual cost of collecting and publishing detailed trade statistics (£10 million approximately) is regarded as excessive by Government ministers, particularly in view of the need to improve the statistics identified in this volume.

January 1989

LIST OF APPENDICES

APPENDIX 1: Basis of the Initial Quarterly Estimates of the Elements of the UK Balance of Payments

Credits by sector	Based wholly or partly on reported quarterly data	Projection based on: Data up to previous quarter	Annual data	Less frequent or no reported data
Services				
Government services	X			
Sea transport				
UK operators	X			
Overseas operators	X			
Civil aviation				
UK airlines	X			
Overseas airlines	X		X	
Travel	X			
Financial and other services				
Insurance				
Lloyds	X			
Companies			X	
Brokers	X			
Banking				X*
Commodity traders and export houses				X
Other brokerage			X	
Consulting engineers, management consultants, surveyors			X	
Process engineers			X	
Telecommunication and postal		X		
Films and TV			X	
North Sea oil services		X		
Land transport	X			
Royalties and services (related)			X	
Royalties and services (unrelated incl. oil)			X	
Students			X	
Diplomats				X
US forces	X			
Miscellaneous			X	X
Total percentage†	56	4	40	

† based on 1986 estimates made in March 1987

* Annual in future

Credits by sector (continued)	Based wholly or partly on reported quarterly data	Projection based on		
		Data up to previous quarter	Annual data	Less frequent or no reported data
Transfers				
Government	X			
Private			X	X
Total percentages†	56		44	
Interest, Profits & Dividends				
Direct investment				
Banks			X	
Insurance			X	
Other	X			
Portfolio investment				
Banks	X			
Lloyds	X			
Financial institutions (other)			X	
Other				X
Other investment				
Bank lending	X			
Other UK residents	X			X
Reserves and other central government	X			
Total percentages†	90		10	

Debits by sector	Based wholly or partly on reported quarterly data	Projection based on		
		Data up to previous quarter	Annual data	Less frequent or no reported data
Services				
Government services	X			
Sea transport				
UK operators	X			
Overseas operators				X
Civil aviation				
UK airlines	X			
Overseas airlines		X		
Travel	X			
Financial and other services				
Telecom and postal		X		
Land transport	X			
Films and TV			X	
North Sea oil services		X		
Royalties and services (related)			X	
Royalties and services (unrelated incl. oil)			X	
Other			X	X
Total percentages†	56	16	28	
Transfers				
Government	X		X	
Private			X	X
Total percentages†	65		35	
Interest, Profits & Dividend				
Direct investment				
Banks			X	
Insurance			X	
Other	X			
Portfolio investment				
Public	X			
Private				X
Other investment				
Banks	X			
Other UK residents	X		X	X
General government	X			
Total percentages†	93		7	

APPENDIX 2: SIC 1980 industry groups

Abbreviated Title Given in Tables	Full Description	SIC 1980
Manufacturing		
Food and tobacco	Food, drink and tobacco	Class 41/42
Chemicals	Chemical and Allied Industries and Coal	Classes 25 and 26
Metal	Metal manufacture	Class 22
Mechanical engineering	Mechanical and instrument engineering	Classes 32 and 37
Electrical engineering	Electrical engineering	Classes 33 and 34
Shipbuilding	Shipbuilding, marine engineering and vehicles other than motor vehicles	Group 361
Motor	Motor vehicle manufacture	Class 35 plus groups 362, 363, 364 and 365
Textiles	Textiles, leather, clothing and footwear	Classes 43, 44 and 45
Paper	Paper, printing and publishing	Class 47
Rubber	Rubber	Class 48
Other manufacture	Metal goods NES, bricks, pottery, glass, cement, timber, furniture and other manufacturing industries	Classes 24, 31, 46 and 49
Total manufacture		
Non-Manufacturing		
Agriculture	Agriculture, forestry and fishing	Division 0
Mining and petroleum	Mining, quarrying and petroleum	Classes 11, 12, 13, 14, 21 and 23
Construction	Construction	Division 5
Transport	Transport and communications including shipping	Division 7
Distribution	Distributive trades	Classes 61, 62, 63, 64 and 65
Other financial	Financial services other than banking and insurance	Group 815
Property	Property owning and managing	Classes 66 and 85
Other activities	Other activities including banking and insurance	Division 9 plus classes 67, 82, 83, 84 Group 814
Total non-manufacture		
Total all industries		

APPENDIX 3: The Economic Significance of Tourism with the European Community

Classification	Description	Coverage	Frequency
Intra-Community Tourism (million ECU)	Receipts in/from each Member State and in Total	EC	Annual
International Tourist Receipts (million ECU) (% GDP)	Receipts by Member State and in Total	EC and Rest of the World	Annual
International Tourist Expenditures (million ECU) (% PFC*)	Expenditure by Member State and in Total	EC and Rest of the World	Annual
International Tourist Receipts – Expenditure as a Percentage of Exports – Imports of Goods and Services	By Member State and in Total	EC and Rest of the World	Annual
Balance on International Tourism compared with the Balance on Goods and Services excluding Tourism (million ECU) (%)	By Member State and in Total	EC and Rest of the World	Annual
Growth of International Tourist Receipts and Expenditures 1972 = 100	Receipts and Expenditures by Member State and in Total Current and Constant Prices	EC and Rest of the World	1972-1981

*PFC (Private Final Consumption)

APPENDIX 4: British Invisible Exports Council – World Invisible Trade Top '35' Countries

Country
UK
United States

Other EEC
Belgium, Luxembourg
Denmark
France
Greece
Italy
Netherlands
Spain
W.Germany

Other W.Europe
Austria
Finland
Norway
Sweden
Switzerland
Yugoslavia

Australia
Canada
Japan
S.Africa

Brazil
Mexico
Panama
Venezuela

Egypt
Israel
Kuwait
Saudi Arabia

India
Republic of Korea
Malaysia
Philippines
Singapore
Taiwan
Thailand

Total Top '35'

APPENDIX 5: Sources of results of inquiries on book values of inward and outward direct investment, 1962 to 1984

1984 Inquiry

See Appendix 7

1981 Inquiry

Business Statistics Office *Business Monitor* MA4, 1981 Supplement, Census of Overseas Assets, 1981, HMSO, [QRL.10]

"Book values of overseas direct investment at end of 1981" *British Business* 2nd March 1984 [QRL.7]

1978 Inquiry

Business Statistics Office *Business Monitor* MA4, 1978 Supplement, Census of Overseas Assets, 1978, HMSO, [QRL.10]

"Book values of overseas direct investment at end of 1978" *British Business* 27th February 1981, pp. 425-430 [QRL.7]

1974 Inquiry

Business Statistics Office *Business Monitor* M4, 1974 Supplement, Census of Overseas Assets, 1974, HMSO, [QRL.10]

"Book values of overseas direct investments" *Trade and Industry* 25th February 1977, pp. 528-535 [QRL.60]

1971 Inquiry

Business Statistics Office *Business Monitor* M4, Part II, 1972, Overseas Transactions, HMSO, [QRL.10]

"Book values of overseas direct investments" *Trade and Industry* 15th November 1973, pp. 368-375 [QRL.60]

1968 Inquiry

Government Statistical Service *Business Monitor* M4, 1969, Overseas Transactions, Tables 34-41, HMSO, [QRL.10]

"Book values of overseas direct investments" *Board of Trade Journal* 23rd September 1970, pp. 644-649 [QRL.5]

1965 Inquiry

"Book values of overseas investments" *Board of Trade Journal* 26th January 1968, pp. i-xxiii [QRL.5]

1962 Inquiry

"Book values of UK investment overseas, and of overseas investments in the United Kingdom" *Board of Trade Journal* 7th August 1964, pp. 292-296 [QRL.5]

"Value of net assets of overseas direct investments" *Board of Trade Journal* 15th November 1963, pp. 1080-1081 [QRL.5]

APPENDIX 6: Sources of results of inquiries on annual transactions of inward and outward Direct Investment

1984 Inquiry

Business Statistics Office *Business Monitor* MA4, Overseas Transactions 1984 HMSO [QRL.10] (Designated as a 'main source' (see 5.3.1.7))

"Overseas Direct Investment in 1984" *British Business* 6th June 1986, pp. 459-465 [QRL.7]

1983 Inquiry

Business Statistics Office *Business Monitor* MA4, Overseas Transactions 1983, HMSO [QRL.10] (Designated as a 'main source' (see 5.3.1.7))

"Overseas Direct Investment Up in 1983" *British Business* 12th May 1985, pp. 305-310 [QRL.7]

"UK Overseas Direct Investment in 1983" *British Business* 16th August 1985, pp. 338-339 [QRL.7]

1982 Inquiry

Business Statistics Office *Business Monitor* MA4, Overseas Transactions, 1982, HMSO [QRL.10] (Designated as a 'main source' (see 5.3.1.7))

"UK Companies Invest a Further £2.5 billion Abroad" *British Business* 18th May 1984, pp. 36-41 [QRL.7]

1981 Inquiry

Business Statistics Office *Business Monitor* MA4, Overseas Transactions, 1981, HMSO, [QRL.10] (Designated as a 'main source' (see 5.3.1.7))

"UK Foreign Investment Up 40 per cent" *British Business* 13th May 1983, pp. 284-290 [QRL.7]

1980 Inquiry

Business Statistics Office *Business Monitor* MA4, Overseas Transactions, 1980, HMSO [QRL.10] (Designated as a 'main source' (see 5.3.1.7))

"Inward Investment Increases to £2.6 billion" *British Business* 14th May 1982, pp. 76-81 [QRL.7]

1979 Inquiry

Business Statistics Office *Business Monitor* MA4, Overseas Transactions, 1979, HMSO [QRL.10] (Designated as a 'main source' (see 5.3.1.7))

"Overseas Net Direct Investment in 1979" *British Business* 15th May 1981, pp. 121-127 [QRL.7]

1978 Inquiry

Business Statistics Office *Business Monitor* MA4, Overseas Transactions, 1978, HMSO [QRL.10] (Designated as a 'main source' (see 5.3.1.7))

"Overseas Net Direct Investment in 1978" *British Business* 6th June 1980, pp. 131-136 [QRL.7]

1977 Inquiry

Business Statistics Office *Business Monitor* MA4, Overseas Transactions, 1977, HMSO [QRL.10] (Designated as a 'main source' (see 5.3.1.7))

"Overseas Direct Investment in 1977" *Trade and Industry* 23rd March 1979, pp. 642-648 [QRL.60]

1976 Inquiry

Business Statistics Office *Business Monitor* M4, Overseas Transactions, 1976, HMSO [QRL.10] (Designated as a 'main source' (see 5.3.1.7))

"Overseas Direct Net Investment in 1976" *Trade and Industry* 9th June 1978, pp. 562-571 [QRL.60]

1975 Inquiry

Business Statistics Office *Business Monitor* M4, Overseas Transactions, 1975, HMSO [QRL.10] (Designated as a 'main source' (see 5.3.1.7))

"Overseas Direct Investment Recovers in 1976" *Trade and Industry* 5th August 1977, pp. 251-254 [QRL.60]

1974 Inquiry

Business Statistics Office *Business Monitor* M4, Overseas Transactions, 1974, HMSO [QRL.10] (Designated as a 'main source' (see 5.3.1.7))

"Overseas Direct Investment in 1974" *Trade and Industry* 25th June 1976, pp. 800-806 [QRL.60]

1973 Inquiry

Business Statistics Office *Business Monitor* M4, Overseas Transactions, 1973, HMSO [QRL.10] (Designated as a 'main source' (see 5.3.1.7))

"Overseas Direct Investment in 1973" *Trade and Industry* 14th March 1975, pp. 705-709 [QRL.60]

"Overseas Direct Investment in 1973" *Trade and Industry* 21st March 1975, pp. 804-805 [QRL.60]

1972 Inquiry

Business Statistics Office *Business Monitor* M4, Part 1, Overseas Transactions, 1972, HMSO [QRL.10] (Designated as a 'main source' (see 5.3.1.7))

"Overseas Direct Investment in 1972" *Trade and Industry* 13th June 1974, pp. 554-558 [QRL.60]

1971 Inquiry

Government Statistical Service *Business Monitor* M4, Overseas Transactions, 1971, HMSO [QRL.10] (Designated as a 'main source' (see 5.3.1.7))

"Overseas Direct Investment in 1971" *Trade and Industry* 29th March 1973, pp. 702-706 [QRL.60]

1970 Inquiry

Government Statistical Service *Business Monitor* M4, Overseas Transactions, 1970, HMSO [QRL.10] (Designated as a 'main source' (see 5.3.1.7))

"Overseas Direct Investment in 1970" *Trade and Industry* 23rd March 1972, pp. 514-518 [QRL.60]

1969 Inquiry

Government Statistical Service *Business Monitor* M4, Overseas Transactions, 1969, HMSO [QRL.10] (Designated as a 'main source' (see 5.3.1.7))

"Overseas Direct Investment in 1969" *Trade and Industry* 7th April 1971, pp. 30-33 [QRL.60]

1968 Inquiry

Government Statistical Service *Business Monitor* M4, Overseas Transactions, 1968, HMSO [QRL.10] (Designated as a 'main source' (see 5.3.1.7))

"Overseas Direct Investment in 1968" *Board of Trade Journal* 8th April 1970, pp. 954-957 [QRL.5]

1967 Inquiry

"Overseas Investment in 1967" *Board of Trade Journal* 9th May 1969, pp. 1300-1318 [QRL.5] (Designated as a 'main source' (see 5.3.1.7))

1966 Inquiry

"Overseas Investment in 1966" *Board of Trade Journal* 19th July, 1968 pp. i-xxiii [QRL.5] (Designated as a 'main source' (see 5.3.1.7))

1965 Inquiry

"Overseas Investment" *Board of Trade Journal* 30th June 1967, pp. i-xxvii [QRL.5] (Designated as a 'main source' (see 5.3.1.7))

1964 Inquiry

"United Kingdom Direct Investment Overseas 1961 to 1965" *Board of Trade Journal* 10th June 1966, pp. 1304-1311 [QRL.5]

1963 Inquiry

"United Kingdom Direct Investment Overseas 1960 to 1964" *Board of Trade Journal* 2nd April 1965, pp. 729-735 [QRL.5]

"UK Private Direct Investment Overseas" *Board of Trade Journal* 10th September 1965, p582 [QRL.5]

1962 Inquiry

"United Kingdom Direct Investment Overseas 1959 to 1963" *Board of Trade Journal* 7th August 1964, pp. 286-291 [QRL.5]

1961 Inquiry

"United Kingdom Direct Investment Overseas 1959 to 1962" *Board of Trade Journal* 15th November 1963, pp. 1082-1086 [QRL.5]

"United Kingdom Direct Investment Overseas, 1958 to 1962" *Board of Trade Journal* 19th April, pp. 877-883 [QRL.5]

1960 Inquiry

"United Kingdom Direct Investment Overseas 1958 to 1961" *Board of Trade Journal* 14th September 1962, pp. 555-559 [QRL.5]

"United Kingdom Direct Investment Overseas 1958 to 1961" *Board of Trade Journal* 30th March 1962, pp. 674-679 [QRL.5]

1958/1959 Inquiries

"United Kingdom Direct Investment Overseas 1958 to 1960" *Board of Trade Journal* 6th October 1961, pp. 715-720 [QRL.5]

"United Kingdom Direct Investment Overseas 1958 to 1959" *Board of Trade Journal* 17th February 1961, pp. 386-390 [QRL.5]

"United Kingdom Direct Investment Overseas 1958 to 1959" *Board of Trade Journal* 7th October 1960, pp. 803-805 [QRL.5]

APPENDIX 7: Preliminary Results of the 1984 Book Value Survey

Following the completion of the main text in this chapter the first results of the 1984 survey of inward direct investment book values became available. These results were included in an article ("Book value of UK net investment overseas at end of 1984") in *British Business* [QRL.7] for 22nd May 1987.

The 1984 data provided in the *British Business* article provides a detailed country breakdown of outward assets (table 2) and inward assets (table 4), for all industries (except property), including oil, banking and insurance. Also 1984 figures, broken down into manufacturing and non-manufacturing and by broad geographical area, are given for outward assets in table 3 and for inward assets in table 5. Table 1 includes aggregate outward and inward asset data for 1984 by broad sector (i.e banks; insurance; oil; other companies).

It is important to note that the oil companies data incorporated in these 1984 figures are derived from a Department of Trade and Industry enquiry, rather than from that of the Bank of England as previously. This, in fact, resulted in changes in the nature of the coverage of oil companies, such that the 1984 figures are not strictly comparable to any available for earlier years. However, bringing the oil companies survey into the DTI's integrated coverage makes the treatment of the industry more in line with the other areas covered.

More detailed coverage of these asset surveys, along the lines of the data scrutinised in 5.2 and 5.4, was scheduled for late 1987 in *Business Monitor MO4, Direct Investment: External Assets and Liabilities 1984*.

APPENDIX 8: Sources of data on royalties surveys

1984 Inquiry

Business Statistics Office *Business Monitor MA4, Overseas Transactions 1984* Tables 5.1 to 5.3 HMSO [QRL.10]

"Overseas earnings from royalties net £85m" *British Business* 15th August, 1986, pp48-49 [QRL.7]

1983 Inquriy

Business Statistics Office *Business Monitor MA4, Overseas Transactions 1983* Tables 5.1 to 5.3 HMSO [QRL.10]

"Overseas earnings from royalties net £155m" *British Business* 24th May 1985, pp357-358 [QRL.7]

1982 Inquiry

Business Statistics Office *Business Monitor MA4, Overseas Transactions 1982* Tables 5.1 to 5.3 HMSO [QRL.10]

"Overseas earnings from royalties net £148m" *British Business* 25th May 1984, pp100-101 [QRL.7]

1981 Inquiry

Business Statistics Office *Business Monitor MA4, Overseas Transactions 1981* Tables 5.1 to 5.5 HMSO [QRL.10]

"Overseas earnings from royalties: £151 net" *British Business* 27th May 1983, pp368-370 [QRL.7]

1980 Inquiry

Business Statistics Office *Business Monitor MA4, Overseas Transactions 1980* Tables 5.1 to 5.5 HMSO [QRL.10]

"Royalties earnings net £119m" *British Business* 28th May 1982, pp190-191 [QRL.7]

1979 Inquiry

Business Statistics Office *Business Monitor MA4, Overseas Transactions 1979* Tables 5.1 to 5.5 HMSO [QRL.10]

"Overseas royalties and similar transactions: 1979" *British Business* 12th June 1981, pp308-310 [QRL.7]

1978 Inquiry

Business Statistics Office *Business Monitor MA4, Overseas Transactions 1978* Tables 5.1 to 5.5 HMSO [QRL.10]

"Overseas royalties and similar transactions in 1978" *British Business* 4th July 1980, pp359-361 [QRL.7]

1977 Inquiry

Business Statistics Office *Business Monitor MA4, Overseas Transactions 1977* Tables 5.1 to 5.5 HMSO [QRL.10]

"Overseas royalties and similar transactions in 1977" *Trade and Industry* 8th June 1979, pp453-455 [QRL.60]

1976 Inquiry

Business Statistics Office *Business Monitor M4, Overseas Transactions 1976* Tables 5.1 to 5.5 HMSO [QRL.10]

"Overseas royalties and similar transactions: net receipts rise to £91m in 1976" *Trade and Industry* 14th July 1978, pp99-101 [QRL.60]

1975 Inquiry

Business Statistics Office *Business Monitor M4, Overseas Transactions 1975* Tables 5.1 to 5.5 HMSO [QRL.10]

"Overseas royalties and similar transactions in 1975" *Trade and Industry* 12th August 1977, pp293-295 [QRL.60]

1974 Inquiry

Business Statistics Office *Business Monitor M4, Overseas Transactions 1974* Tables 37 to 41 HMSO [QRL.10]

"Overseas royalties transactions in 1974" *Trade and Industry* 25th June 1976, pp804-806 [QRL.60]

1973 Inquiry

Business Statistics Office *Business Monitor M4, Overseas Transactions 1973* Tables 37 to 41 HMSO [QRL.10]

"Overseas royalties transactions in 1973" *Trade and Industry* 11th April 1975, pp104-106 [QRL.60]

1972 Inquiry

Business Statistics Office *Business Monitor M4, Part II, Overseas Transactions 1972* Tables 50 to 54 HMSO [QRL.10]

"Overseas royalties transactions in 1972" *Trade and Industry* 27th June 1974, pp653-654 [QRL.60]

1971 Inquiry

"Overseas royalty transactions in 1971" *Trade and Industry* 12th April 1973, pp105-107 [QRL.60]

1970 Inquiry

"Overseas royalty transactions in 1970" *Trade and Industry* 30th March 1972, pp575-577 [QRL.60]

1969 Inquiry

"Overseas royalty transactions in 1969" *Trade and Industry* 31st March 1971, pp701-703 [QRL.60]

1968 Inquiry
"Overseas royalty transactions in 1968" *Board of Trade Journal* 25th March 1970, pp839-841 [QRL.5]

1967 Inquiry
"Overseas royalty transactions in 1967" *Board of Trade Journal* 14th May 1969, pp1392-1394 [QRL.5]

1966 Inquiry
"Overseas royalty transactions 1966" *Board of Trade Journal* 23rd August 1968, pp548-550 [QRL.5]

1965 Inquiry
"Overseas transactions-trade credit and royalties" *Board of Trade Journal* 21st July 1967, pp160-167 [QRL.5]

1964 Inquiry
"Overseas royalty transactions in 1964" *Board of Trade Journal* 29th July 1966, pp 276-279 [QRL.5]

APPENDIX 9: Sources of data on Films and TV surveys

1985 Inquiry

"British films and television earn £164m overseas in 1985" *British Business* 19th September 1986, pp41-42 [QRL.7]

1984 Inquiry

Business Statistics Office *Business Monitor MA4, Overseas Transactions 1984* Table 6.1 to 6.4 HMSO [QRL.10]

"British films and TV earn £125m overseas in 1984" *British Business* 30th August 1985, pp430-432 [QRL.7]

1983 Inquiry

Business Statistics Office *Business Monitor MA4, Overseas Transactions 1983* Table 6.1 to 6.4 HMSO [QRL.10]

"British films and TV earned £101m overseas last year" *British Business* 5th October 1984, pp200-202 [QRL.7]

1982 Inquiry

Business Statistics Office *Business Monitor MA4, Overseas Transactions 1982* Tables 6.1 to 6.4 HMSO [QRL.10]

"British films and television earned £78m overseas last year" *British Business* 16th September 1983, pp200-201 [QRL.7]

1981 Inquiry

Business Statistics Office *Business Monitor MA4, Overseas Transactions 1981* Tables 7.1 to 7.5 HMSO [QRL.10]

"British films and television earned £41 million overseas last year" *British Business* 17th September 1982, pp72-73 [QRL.7]

1980 Inquiry

Business Statistics Office *Business Monitor MA4, Overseas Transactions 1980* Tables 7.1 to 7.5 HMSO [QRL.10]

"Film and television overseas earnings reach new high" *British Business* 4th September 1981, pp19-20 [QRL.7]

1979 Inquiry

Business Statistics Office *Business Monitor MA4, Overseas Transactions 1979* Tables 7.1 to 7.5 HMSO [QRL.10]

"Overseas transactions in films and television material in 1979" *Trade and Industry* 19th December 1980, pp794-795 [QRL.60]

1978 Inquiry

Business Statistics Office *Business Monitor MA4, Overseas Transactions 1978* Tables 7.1 to 7.5 HMSO [QRL.10]

"Films and television material: overseas transactions in 1978" *Trade and Industry* 6th July 1979, pp52-53 [QRL.60]

1977 Inquiry

Business Statistics Office *Business Monitor MA4, Overseas Transactions 1977* Tables 7.1 to 7.5 HMSO [QRL.10]

"Films and television material: overseas transactions in 1977" *Trade and Industry* 25th August 1978, pp472-473 [QRL.60]

1976 Inquiry

Business Statistics Office *Business Monitor MA, Overseas Transactions 1976* Tables 7.1 to 7.5 HMSO [QRL.10]

"Films and television material: overseas transactions in 1976" *Trade and Industry* 26th August 1977, pp379-380 [QRL.60]

1975 Inquiry

Business Statistics Office *Business Monitor M4, Overseas Transactions 1975* Tables 7.1 to 7.5 HMSO [QRL.10]

"Film and television material: overseas transactions in 1975" *Trade and Industry* 20th August 1976, pp534-536 [QRL.60]

1974 Inquiry

Business Statistics Office *Business Monitor M4, Overseas Transactions 1974* Tables 48 to 52 HMSO [QRL.10]

"Film and television material: overseas transactions in 1974" *Trade and Industry* 8th August 1975, pp362-364 [QRL.60]

1973 Inquiry

Business Statistics Office *Business Monitor M4, Overseas Transactions 1973* Tables 48-52 HMSO [QRL.10]

"Overseas transactions in 1973. Film and television material" *Trade and Industry* 1st August 1974, pp242-244 [QRL.60]

1972 Inquiry

"Overseas transactions in 1972. Film and television material" *Trade and Industry* 9th August 1973, pp318-319 [QRL.60]

1971 Inquiry

"Overseas transactions in 1971: Film and television material" *Trade and Industry* 20th July 1972, pp128-129 [QRL.60]

1970 Inquiry

"Overseas transactions in 1970: film and television material" *Trade and Industry* 5th August 1971, pp241-243 [QRL.60]

1969 Inquiry

"Overseas transactions in 1969: film and television material" *Board of Trade Journal* 22nd July 1970, pp170-171 [QRL.5]

1968 Inquiry

"Overseas transactions in 1968: film and television material" *Board of Trade Journal* 5th November 1969, pp1236-1238 [QRL.5]

1967 Inquiry

"Overseas transactions 1967: film and television material" *Board of Trade Journal* 6th December 1968, pp1540-1542 [QRL.5]

1966 Inquiry

"Overseas transactions: films and television material" *Board of Trade Journal* 3rd May 1968, pp1360-1362 [QRL.5]

1965 Inquiry

"Overseas transactions: films and telelvision material" *Board of Trade Journal* 6th January 1967, pp22-24 [QRL.5]

1964 Inquiry

"Overseas payments and earnings for cinematograph films" *Board of Trade Journal* 31st December 1965, p1568 [QRL.5]

1963 Inquiry

"Overseas payments and earnings for cinematograph films" *Board of Trade Journal* 25th September 1964, p688 [QRL.5]

APPENDIX 10: Sources of data on overseas transactions inquiry export analysis

1981 Inquiry

Business Statistics Office *Business Monitor MA4, Overseas Transactions 1981* Tables 6.1 to 6.4 HMSO [QRL.10]

"Company analysis of direct exporters in 1981" *British Business* 27th May 1983, pp366-367 [QRL.7]

1980 Inquiry

Business Statistics Office *Business Monitor MA4, Overseas Transactions 1980* Tables 6.1 to 6.4 HMSO [QRL.10]

"One hundred and one firms account for half UK exports" *British Business* 28th May 1982, pp192-193 [QRL.7]

1979 Inquiry

Business Statistics Office *Business Monitor MA4, Overseas Transactions 1979* Tables 6.1 to 6.4 HMSO [QRL.10]

"Ninety-four firms account for half Britain's exports" *British Business* 3rd July 1981, pp447-448 [QRL.7]

1978 Inquiry

Business Statistics Office *Business Monitor MA4, Overseas Transactions 1978* Tables 6.1 to 6.6 HMSO [QRL.10]

"Direct exporters and credit terms of exports in 1978" *British Business* 20th June 1980, pp236-237 [QRL.7]

1977 Inquiry

Business Statistics Office *Business Monitor MA4, Overseas Transactions 1977* Tables 6.1 to 6.6 HMSO [QRL.10]

"Direct exporters and credit terms of exports in 1977" *Trade and Industry* 20th July 1979, pp144-145 [QRL.60]

1976 Inquiry

Business Statistics Office *Business Monitor M4, Overseas Transactions 1976* Tables 6.1 to 6.6 HMSO [QRL.10]

"Direct exporters and credit terms of exports in 1976" *Trade and Industry* 21st July 1978, pp162-163 [QRL.60]

1975 Inquiry

Business Statistics Office *Business Monitor M4, Overseas Transactions 1975* Tables 6.1 to 6.6 HMSO [QRL.10]

"Direct exporters and the credit terms of exports" *Trade and Industry* 12th August 1977, pp291-292 [QRL.60]

1974 Inquiry

Business Statistics Office *Business Monitor M4, Overseas Transactions 1974* Tables 42 to 47 HMSO [QRL.10]

"Direct exporters and the credit terms of exports" *Trade and Industry* 16th July 1976, pp202-203 [QRL.60]

1973 Inquiry

Business Statistics Office *Business Monitor M4, Overseas Transactions 1973* Tables 42 to 47 HMSO [QRL.10]

"Direct exporters and the credit terms of exports" *Trade and Industry* 11th April 1975, pp101-103 [QRL.60]

1972 Inquiry

Business Statistics Office *Business Monitor M4 Part II, Overseas Transactions 1972* Tables 55 to 60 HMSO [QRL.10]

"Direct exporters and the credit terms of exports" *Trade and Industry* 27th June 1974, pp651-652 [QRL.60]

1971 Inquiry

"Direct exporters and the credit terms of exports" *Trade and Industry* 10th May 1973, pp272-274 [QRL.60]

1970 Inquiry

"Direct exporters and the credit terms of exports" *Trade and Industry* 13th April 1972, pp50-52 [QRL.60]

1969 Inquiry

"Trade credit and exports in 1969" *Trade and Industry* 14th April 1971, pp91-93 [QRL.60]

1968 Inquiry

"Trade credit and exports" *Board of Trade Journal* 15th April 1970, pp1020-1022 [QRL.5]

1967 Inquiry

"Trade credit and exports in 1967" *Board of Trade Journal* 3rd September 1969, pp614-617 [QRL.5]

1966 Inquiry

"Overseas transactions in 1966 –trade credit and exports" *Board of Trade Journal* 16th August 1968, pp468-475 [QRL.5]

1965 Inquiry

"Size distribution of exporting firms" *Board of Trade Journal* 30th June 1967, p xxviii [QRL.5]

APPENDIX 11: Specimen Forms

1. Annual inquiry into transactions of UK companies with their overseas subsidiaries, associates amd branches in 1986
2. Overseas assets and liabilities at end 1984 of overseas subsidiaries, branches and associates of UK companies.
3. Annual inquiry into transactions of UK oil companies with their overseas subsidiaries, associates and branches in 1986.
4. Assets and liabilities at end 1986 of overseas subsidiaries, branches and associates of UK oil companies.
5. Annual inquiry into overseas transactions of oil companies operating onshore in the UK in 1986.
6. Assets and liabilities at end 1986 of overseas oil companies operating in the UK.
7. Annual inquiry into overseas transactions of oil companies having interests in the UKCS in 1986.
8. Assets and liabilities at end 1986 of overseas oil companies with interests in the UK sector of the continental shelf
9. Annual inquiry into transactions between UK branches of oil companies and their overseas head office in 1986.
10. Assets and liabilities at end 1986 of UK branches of overseas oil companies.
11. Annual inquiry into transactions of UK companies with their overseas direct investors in 1986.
12. Annual inquiry into transactions of UK branches with their overseas head office in 1986.
13. Assets and liabilities at end 1984 of UK subsidiaries and associates of overseas companies.
14. Assets and liabilities at end 1984 of UK branches of overseas companies.

A **compulsory** inquiry conducted by the Government Statistical Service
IN CONFIDENCE

DEPARTMENT OF TRADE AND INDUSTRY
Statistics Division 2
Room 252, Sanctuary Buildings
20 Great Smith Street
LONDON SW1P 3DB

Telepnone 01-215

Date. 1 April 1987

	Ref	Sub	YE	UKInd	Tp	Roy	Co
1		86					

OT/OUT/86

← Please quote this reference number in any correspondence

Please correct any errors in name or address

NOTICE UNDER SECTION 1 OF THE STATISTICS OF TRADE ACT 1947

The Secretary of State for Trade and Industry hereby requires you by law to provide the information called for in this inquiry form with the exception of questions 6,7,11 12. The information requested in this form is required to compile the balance of payments, for the appreciation of economic trends and for the provision of a statistical service for government and industry.

Please complete this form and return it to the above address by 1 September 1987

N.B. The information given by you under the Act will be treated as confidential in strict accordance with the Act and subject to the further restriction that information about individual businesses will be used and disclosed under Ministerial direction to other Government departments for statistical purposes only, except that the names and addresses of individual businesses and their industrial classification may, if a Minister so directs, be made available to Ministers and officials of government departments who need them for carrying out their functions

ANNUAL INQUIRY INTO TRANSACTIONS OF UNITED KINGDOM COMPANIES WITH THEIR OVERSEAS SUBSIDIARIES, ASSOCIATES AND BRANCHES IN 1986

Dear Contributor

We are conducting this annual inquiry to obtain up-to-date statistical information about various overseas transactions and earnings of United Kingdom companies for use in compiling our country's balance of payments statistics. The information requested is the type of statistics which any government needs in deciding its economic policy as well as its commercial policy in relation to other countries. This inquiry is the only source of such information and without it the accuracy of the UK balance of payments statistics will be greatly reduced.

Please note that questions 6, 7, 11 and 12 are voluntary and you are not required by law to answer them. However it would be very helpful if you could give this information as it will enable us to provide a fuller analysis of the returns. Any information given voluntarily will be treated as confidential. It will not be disclosed outside this office and will be used only for compiling totals which will not disclose the figures for individual contributors.

The form should be returned even where there is little or nothing to report. We are prepared to accept your return in the form of a computer printout if you prefer - please see the introduction of the enclosed notes.

A summary of the results will be published in British business. More detailed results will be published later in the Business Monitor, MA4, Overseas Transactions, 1986.

If you have any difficulties or require more information about the inquiry my staff will be pleased to help you. The telephone extension of the person dealing with this inquiry is shown above.

I would be grateful if you could return the completed form to us as soon as possible. **In any case it must be returned by 1 September 1987.**

Yours faithfully

P H Richardson.

P. H. RICHARDSON
Chief Statistician

SC107A

IMPORTANT Please refer to the enclosed notes before completing this form. If you do not have precise figures readily available give the best estimates you can. Please show all values in sterling rounded in £ million to one decimal place. (e.g. £53,400,000 should be entered as 53.4). Use the same rates of conversion as in your own accounts.

1. **DATE OF FINANCIAL YEAR END** if not **31 December 1986** (see note 1) .... / /

PART A – OVERSEAS SUBSIDIARY AND ASSOCIATE COMPANIES

2. **GENERAL PARTICULARS** (see note 2)

2.1 Country in which overseas subsidiary or associate is incorporated ....

2.2 Country of ownership of overseas subsidiary or associate
If company is directly owned by UK reporting group or company write "UK" in the box if indirectly owned through another overseas company give country of that company directly owning the subsidiary or associate.

3. **AMOUNTS RECEIVED FROM AND PAID TO OVERSEAS SUBSIDIARIES AND ASSOCIATES** by UK reporting group or company

3.1 **Interest received** net of any withholding tax paid overseas, less interest paid net of any UK withholding tax you deducted during the year. (If payments exceed receipts enter figures as a negative.) ....

3.2 **Dividends received** (see note 3.2). ....

3.3 **Royalties and similar transactions** on revenue and capital account (see notes 3.3 and 3.4)

a. Total amount received from overseas company less any tax paid overseas ....

b. Total amount paid to overseas company less any UK tax you deducted ....

3.4 **All other transactions not elsewhere specified** (see notes 3.3 and 3.5)

a. Total amount received from overseas company less any tax paid overseas ....

b. Total amount paid to overseas company less any UK tax you deducted ....

4. **DEALINGS IN THE OVERSEAS COMPANIES' SHARES, DEBENTURES, AND QUOTED LOAN STOCK** (see note 4)

4.1 Purchases (including new subscriptions) ....

4.2 Sales (including proceeds from redemptions) ....

5. **NET TOTAL OF INTER-COMPANY CURRENT ACCOUNT BALANCES AND LOANS** (see notes 4b and 5)

5.1 At 31.12.86 Net amount due from UK reporting group or company ....
or
Net amount due to UK reporting group or company ....

5.2 At 31.12.85 Net amount due from UK reporting group or company ....
or
Net amount due to UK reporting group or company ....

6. **UK REPORTING GROUP'S OR COMPANY'S SHARE OF OVERSEAS COMPANY'S PROFIT OR LOSS** (whether remitted or not) (see note 6)
This question is voluntary.

6.1 Profit ....
or
6.2 Loss ....

7. **NET GAIN OR LOSS RESULTING FROM FOREIGN EXCHANGE RATE MOVEMENTS** made by the UK reporting group or company on any transactions with the overseas concern (see note 7).
This question is voluntary.

7.1 Gain ....
or
7.2 Loss ....

Please also complete Part B overleaf if appropriate

OT/OUT/86

For official use

	Ref	Sub	Form
5			86

	1	2	3	4	5	6	7	8
61								
62								
60								

	TOTAL (see note 2.1)								
	£ million	£ million	£ million	£ million	£ million	£ million	£ million	£ million	£ million
65									
66									
103									
104									
113									
114									
73									
74									
76									
77									
78									
79									
63									
64									
67									
68									

PLEASE COMPLETE THIS FORM AND RETURN IT BY 1 SEPTEMBER 1987

SC107C

For official use

	Ref	Sub		Form
6			86	

	1	2	3
21			
22			

PART B – OVERSEAS BRANCHES

8. GENERAL PARTICULARS (see notes 2.1 and 2.2)

Country in which overseas branch is situated .... | | | |

		TOTAL (see note 2.1) £ million	£ million	£ million	£ million
9. AMOUNTS RECEIVED FROM AND PAID TO OVERSEAS BRANCHES					
9.1 **Royalties and similar transactions** on revenue and capital account (see notes 3.3 and 3.4)					
a. Total amount received from overseas branch less any tax paid overseas	53				
b. Total amount paid to overseas branch less any UK tax you deducted	54				
9.2 **All other transactions not elsewhere specified** (see notes 3.3 and 3.5)					
a. Total amount received from overseas branch less any tax paid overseas	57				
b. Total amount paid to overseas branch less any UK tax you deducted	58				
10. NET ASSETS OF OVERSEAS BRANCH ATTRIBUTABLE TO UK REPORTING GROUP OR COMPANY (see note 10)					
10.1 At 31.12.86 Net amount due from UK reporting group or company	26				
or Net amount due to UK reporting group or company	27				
10.2 At 31.12.85 Net amount due from UK reporting group or company	28				
or Net amount due to UK reporting group or company	29				
11. TOTAL PROFIT OR LOSS OF OVERSEAS BRANCH ATTRIBUTABLE TO UK REPORTING GROUP OR COMPANY (whether remitted or not) (see note 11) This question is voluntary.					
11.1 Profit	23				
or 11.2 Loss	24				
12. NET GAIN OR LOSS RESULTING FROM FOREIGN EXCHANGE RATE MOVEMENTS made by the UK reporting group or company on any transactions with the overseas branch (see note 7) This question is voluntary.					
12.1 Gain	71				
or 12.2 Loss	72				

Name of person to be contacted should this be necessary ------------------------------
BLOCK CAPITALS PLEASE

Position in company ------------------------------ Signature ----------------

Telephone No/Ext ------------------------------ Telex ---------- Date -------

4 SC107D

OT/OUT/86N

ANNUAL INQUIRY INTO TRANSACTIONS OF UNITED KINGDOM COMPANIES WITH THEIR OVERSEAS SUBSIDIARIES, ASSOCIATES AND BRANCHES IN 1986

NOTES ON COMPLETING FORM OT/OUT/86

INTRODUCTION

UK parent companies should complete this return on behalf of all UK companies within their group

Returns may be made in the form of a computer printout. The printout should be in the same format and order as the questionnaire, with each item identified by the identification number given against that line For example, the Number for question 3.1 is 65, for 3.2 is 66, etc Space should be left on the printout for us to insert lines 61,62 and 60 at the top of Part A, and lines 21 and 22 at the top of Part B. If there are any difficulties please contact us

DEFINITIONS

For the purpose of this return.

a. the definition applied to a **subsidiary** is that given in Section 736 of the Companies Act 1985.

b. an **associate company** is an overseas company of which the UK reporting group or company directly owns part of the equity capital and which is an associated company of the UK reporting group or company as defined in the Statement of Standard Accounting Practice No 1 (SSAP1).

c a **branch** means a permanent establishment as defined for the purpose of the UK tax and UK double taxation agreements, being a fixed place of business overseas through which the business of the UK enterprise is wholly or partly carried on, and which is not a subsidiary. Include as branches joint ventures carried out overseas with UK or overseas companies if these are not conducted through an overseas subsidiary or associate company. Companies incorporated in the UK but operating wholly or mainly abroad should regard their permanent establishments abroad as overseas branches.
A UK incorporated company with its head office overseas should regard its overseas head office as an overseas branch and its UK registered office as its UK head office and complete Part B of the form accordingly.

d **United Kingdom** covers Great Britain and Northern Ireland. Overseas subsidiaries, associates and branches are those situated outside of Great Britain, Northern Ireland, Isle of Man, and the Channel Islands.

VALUATION

Show values in sterling as entered in the UK reporting group's or company's accounts in £ million rounded to one decimal place. If precise figures are not readily available please give the best estimates you can. Where conversion from an overseas currency is involved, please use the same rate of exchange as in your company's own accounts.

NOTES ON INDIVIDUAL QUESTIONS

1. DATE OF FINANCIAL YEAR END

Where possible give details for 31 December 1986. If details are not readily available please give values for your accounting period ending after 30 June 1986 but before 1 July 1987, stating the date to which the figures relate.

2. GENERAL PARTICULARS - QUESTIONS 2 AND 8

2.1 Please enter the total of all overseas direct investments against the relevant headings. If the investments in individual countries are less than £0.1 million these are the only figures required. Otherwise give details relating to each country in which you have a subsidiary, associate or branch in a separate column. The sum of these (allowing for investments of under £0.1 million) should equal the amounts given under "TOTAL". If you use a continuation sheet, only complete the "TOTAL" column on the first sheet.

2.2 If you have more than one subsidiary, associate or branch in a country you can either give combined figures for that country or, if it is easier for you, separate figures for each subsidiary, associate or branch.

2.3 If you have subsidiary companies incorporated in the UK please complete this questionnaire on behalf of all UK companies within your group. If this is not possible please complete this return in respect of your company and arrange for your UK subsidiaries to make separate returns. If you have an overseas subsidiary which is held indirectly through another overseas subsidiary you may either give consolidated figures for each directly owned company covering that company and all its subsidiaries and associates even if some of them are in different countries, or you may give separate figures for each directly and indirectly owned subsidiary. Include in your return any subsidiaries you have established overseas to raise funds on the international capital market. Please let us know if you need additional forms.

RECEIPTS AND PAYMENTS - QUESTIONS 3 AND 9

3.2 State dividends received from the overseas company less any withholding tax deducted overseas, plus any refunds of this withholding tax on current and previous years' dividends received in the year.
Include dividends on preference shares and any tax refunds received directly from overseas revenue authorities.
State dividends received in the year including any declared in earlier years and paid this year. Do not include any dividends declared but not received in the year.

3.3 If it is not possible to split a payment or receipt between royalties and other transactions (e.g. where amounts include both royalties and other services such as advertising, market research, use of distribution and marketing facilities, etc.) please state the total amount in question 3.3 or 9.1 if royalties account for half or more, or in question 3.4 or 9.2 if royalties account for less than half.

3.4 For royalties in questions 3.3 and 9.1

INCLUDE royalties on printed matter, sound recordings, performing rights, patents, licences, trade marks, designs, copyright, mining and quarrying mineral rights, research and development, technical know-how and assistance,

lump-sum royalty settlements made in year,

amounts payable to overseas agents for charges (less discounts) for services rendered in connection with royalties and similar transactions listed above,

total value of royalties etc. which you collect from, or pay to, overseas residents even though you may be acting as an agent.

EXCLUDE royalties etc. included in the invoiced price of goods you bought or sold,

royalties, rentals, etc. in respect of cinematograph films and television programme material.

3.5 For other transactions in questions 3.4 and 9.2
INCLUDE management and head office expenses, directors' fees, agents' commission, advertising, insurance premiums and claims, profit on third country trade, etc.
EXCLUDE cinematograph films and television programme material, UK imports and exports and their shipping and freight costs.

4. DEALINGS IN THE OVERSEAS COMPANIES' SHARES, DEBENTURES, AND QUOTED LOAN STOCK

a. Give cash values not nominal values.

b. INCLUDE acquisition of shares, debentures or quoted loan stock arising from the capitalisation of inter-company current account balances or loans in question 4.1, and show a corresponding fall in the net amount due to you in question 5.1.

c. INCLUDE transactions in quoted and unquoted shares (including preference shares) and debentures of the overseas company, plus any transactions in its loan stock quoted on a stock exchange in the UK or overseas,

acquisitions made in return for equipment, manufacturing rights, securities, etc. supplied by the UK companies to overseas companies,

purchases made by issuing shares of your UK group of companies in exchange for the equity of an overseas company - value this, if possible, at the market value on the day of the acquisition of the shares issued.

EXCLUDE bonus issues of shares arising from the capitalisation of reserves.
unquoted loan stock - include this in question 5.

5. NET TOTAL OF INTER-COMPANY CURRENT ACCOUNT BALANCES AND LOANS

State the net total of inter-company current account balances, long-term loans (other than debentures and quoted loan stock) and short-term loans outstanding between the overseas company and the UK reporting group or company.
INCLUDE unquoted loan stock.
EXCLUDE any dividends declared but not received in the year.

6. UK REPORTING GROUP'S OR COMPANY'S SHARE IN OVERSEAS COMPANY'S PROFIT OR LOSS

The share of the profits of overseas subsidiary and associate companies should be determined in accordance with the UK reporting group or company's proportionate holding of the subsidiary or associate company's equity share capital. Make allowances for depreciation, all overseas tax and extraordinary items. Deduct withholding taxes paid overseas on dividends, net of any refunds of this tax received in the year, including any refunds received directly from overseas revenue authorities.

7. NET GAIN OR LOSS RESULTING FROM FOREIGN EXCHANGE RATE MOVEMENTS - QUESTIONS 7 AND 12

INCLUDE gains or losses resulting from foreign currency exchange rate movements on inter-company current account balances, long-term loans (other than debentures and quoted loan stock) and short-term loans given in question 5, and for overseas branches on the net assets of the branch given in question 10 arising on amounts outstanding at the beginning or end of the year.

If these figures are not available please estimate or if only the total for all countries is available for question 7 write "TOTAL =" across first two columns. If no estimate is possible write "N/A". If zero, please write "NIL".

10. NET ASSETS OF OVERSEAS BRANCH ATTRIBUTABLE TO UK REPORTING GROUP OR COMPANY

State the net assets as shown in the accounts of the UK reporting group or company. These consist of its fixed assets (net of depreciation), its investments and its net current assets which are not due from or to the UK reporting group or company, less long-term loans and other liabilities held by the branch other than those due to the UK reporting group or company.
INCLUDE loans and current account balances of branch due from or to other UK companies in your group.

11. TOTAL PROFIT OR LOSS OF OVERSEAS BRANCH ATTRIBUTABLE TO UK REPORTING GROUP OR COMPANY

Make allowances for depreciation, all overseas tax and extraordinary items.
Deduct withholding taxes paid overseas, net of any refunds.
INCLUDE in profits any interest paid by branch to reporting company or other UK companies in your group.

PLEASE KEEP THESE NOTES FOR REFERENCE

SC107F

A **compulsory** inquiry conducted by the Government Statistical Service

IN CONFIDENCE

DEPARTMENT OF TRADE AND INDUSTRY
Statistics Division 2
Room 252, Sanctuary Buildings
20 Great Smith Street
LONDON SW1P 3DB

Telephone: 01-215-

1 April 1985

CA/OUT/84

Please correct any errors in name or address

NOTICE UNDER SECTION 1 OF THE STATISTICS OF TRADE ACT 1947

The Secretary of State for Trade and Industry hereby requires you by law to provide the information called for in this inquiry form with the exception of questions 4.5/6/7 and 8.3/4. The information in this form is required for the purposes of the appreciation of economic trends and the provision of a statistical service for government and industry.

Please complete this form and return it to the Department of Trade and Industry by 2 September 1985.

NB The information given by you under the Act will be treated as confidential in strict accordance with the Act and subject to the further restriction that information about individual businesses will be used and disclosed under Ministerial direction to other government departments for statistical purposes only, except that the names and addresses of individual businesses and their industrial classification may if a Minister so directs be made available to Ministers and officials of government departments who need them for carrying out their functions

OVERSEAS ASSETS AND LIABILITIES AT END 1984 OF OVERSEAS SUBSIDIARIES, BRANCHES AND ASSOCIATES OF UK COMPANIES

Dear Contributor

We are conducting this three-yearly inquiry to obtain information on the value of United Kingdom companies' investment in their overseas subsidiaries, branches and associates It is the only source of such data and is used in compiling the UK balance of payments, and in deciding government policy in relation to other countries

A summary of the results will be published in British business More detailed results will be published later in the Business Monitor, MA4, Overseas Transactions, 1984 Supplement

Please note that questions 4 5/6/7 and 8 3/4 are voluntary and you are not required by law to answer them However it would be very helpful if you could give this information as it will complete the balance sheets for the overseas subsidiaries or branches Any information given voluntarily will be treated as confidential It will not be disclosed outside this office and will be used only for compiling totals which will not disclose the figures for individual contributors

If the total value of your overseas investments in all countries is between minus £50,000 and plus £50,000 just tick the box indicated, sign and return the form to us You do not have to answer any other questions

This form should be returned even where there is little or nothing to report against some of the questions I should like to emphasise that if you do not have precise figures readily available we are happy to accept your best estimates We are prepared to accept your return in the form of a computer printout if you prefer – please see introduction on the enclosed notes

I would be grateful if you could return the completed form to us as soon as possible In any case it must be returned by 2 September 1985

If you have any difficulties my staff will be please to help you The telephone extension of the person dealing with this inquiry is shown above

Yours faithfully

P H Richardson

P H RICHARDSON
Chief Statistician

IMPORTANT **Please refer to the enclosed notes before completing this form.**
If you do not have precise figures readily available give the best estimates you can.
Please show all values in sterling as entered in the UK reporting group's or company's accounts rounded to the nearest £ thousand (e.g. £53,400 should be entered as 53).

If you and your UK subsidiaries have no overseas subsidiaries, branches or associates tick this box ☐ sign and return the form. You do not have to answer any other questions.

If the total value of your investments in all countries combined (questions 4.2 + 4.4 + 5.2 – 5.1 + 8.4 + 10) is between minus £50,000 and plus £50,000 tick this box ☐ sign and return the form. You do not have to answer any other questions.

1. DATE OF RETURN if not 31 December 1984 (see note 1) . . . . ☐

PART A – OVERSEAS SUBSIDIARY COMPANIES (see note 2)

2. GENERAL PARTICULARS

2.1 Country in which overseas subsidiary is incorporated . . . .

2.2 Main activity of overseas subsidiary . . . .

2.3 Country of Ownership of Overseas Subsidiary (see notes 2b and 2c) . . . .

If company is directly owned by UK reporting group or company write "UK" in box; if indirectly owned through another overseas subsidiary give country of that subsidiary directly owning the company.

2.4 Percentage of overseas subsidiary's share capital and reserves including preference shares, attributable to UK reporting group or company (see note 2d). (You do not have to answer this question if you answer question 4.5 below.) . . . .

3. ASSETS

3.1 Fixed assets after provision for depreciation plus investments in securities (see note 3a) . . . .

3.2 Value of investment by overseas subsidiary in its subsidiaries for which separate figures are given elsewhere in this return (see note 3) (see note 2c for alternative treatment of these subsidiaries) . . . .

3.3 Current assets (see note 3c) . . . .

3.4 **TOTAL ASSETS** (sum of 3.1 to 3.3) . . . .

4. LIABILITIES

4.1 Current liabilities (see note 4a) . . . .

4.2 Long-term liabilities owing to UK reporting group or company (see note 4b) . . . .

4.3 Long-term liabilities other than to UK reporting group or company (see note 4b) . . . .

4.4 Amount of share capital, reserves, and preference shares of overseas subsidiary attributable to the UK reporting group or company . . . .

4.5 Amount of share capital, reserves, including preference shares, of overseas subsidiary attributable to other shareholders (see note 4c) (This question is voluntary) . . . .

4.6 Minority shareholders' interests in any of the overseas company's subsidiaries consolidated in the above figures (see note 4c) (This question is voluntary) . . . .

4.7 Provisions, deferred tax reserves, and any other liabilities (see note 4c) (This question is voluntary) . . . .

4.8 **TOTAL LIABILITIES** (sum of 4.1 to 4.7) (Should equal total assets at 3.4) . . . .

5. ADDITIONAL ITEMS

Inter-company account balances with UK reporting group or company included in

5.1 Current assets at 3.3 above . . . .

5.2 Current liabilities at 4.1 above . . . .

For official use	*	1984	14	Ref	Sub	Type	Y E.	Ind	Size	A	B	C	Amend

501										
502										
505										

504	%	%	%	%	%	%	%	%	%	%

	£ thousand	£ thousand	£ thousand	£ thousand	£ thousand	£ thousand	£ thousand	£ thousand	£ thousand	£ thousand
511										
512										
513										
506										
514										
515										
516										
535										
520										
510										
517										
507										
518										
519										

For official use					
	539				
	540				

PART B – OVERSEAS BRANCHES (see notes 2a and 2b)

6. GENERAL PARTICULARS

6 1 Country in which overseas branch is situated				
6 2 Main activity of overseas branch				

7. ASSETS

		£ thousand	£ thousand	£ thousand	£ thousand
7 1 Fixed assets after provision for depreciation plus investments in securities (see note 3a)	521				
7 2 Current assets (see note 3c)	523				
7 3 **TOTAL ASSETS** (7 1 plus 7.2)	542				

8. LIABILITIES

8 1 Current liabilities excluding amounts due to UK reporting group or company (see note 4a)	524				
8 2 Long-term liabilities other than to UK reporting group or company (see note 4b)	525				
8 3 Provisions, deferred tax reserves, and any other liabilities (see note 4c) (This question is voluntary)	526				
8 4 Net worth of overseas branch attributable to UK reporting group or company (see notes 4c and 4d) (This question is voluntary)	536				
8 5 **TOTAL LIABILITIES** (sum of 8 1 to 8 4) (Should equal total assets at 7 3 above)	543				
For official use	533				
	534				

PART C – OVERSEAS ASSOCIATES (see notes 2a and 2b)

Include only overseas associates directly owned by UK reporting group or company

9. GENERAL PARTICULARS

9 1 Country in which overseas associate company is situated					
9 2 Main activity of overseas associate					
9 3 Percentage of issued share capital owned by the UK reporting group or company (see note 2d)	541	%	%	%	%

10. VALUE OF INVESTMENT

		£ thousand	£ thousand	£ thousand	£ thousand
Value of investment in overseas associate as valued under SSAP1 in the accounts of the UK reporting group or company plus loan and net current account balances due from the overseas associate	527				

Name of person to be contacted should this be necessary

BLOCK CAPITALS PLEASE

Position in company | Signature

Telephone No/Ext | Telex | Date

OVERSEAS ASSETS AND LIABILITIES AT END 1984 OF OVERSEAS SUBSIDIARIES, BRANCHES AND ASSOCIATES OF UK COMPANIES

NOTES ON COMPLETING FORM CA/OUT/84

INTRODUCTION

UK parent companies should complete this return on behalf of all UK companies within their group, unless alternative arrangements were made for the previous inquiry For groups which chose in the last inquiry to have their subsidiary UK companies complete separate returns, separate forms have been sent to those subsidiary UK companies, and the parent company has been notified about this If any other group wishes to make similar arrangements, the parent company should make its own returns (where applicable) and arrange for its subsidiary UK companies to make separate returns. The necessary additional forms will be supplied on request

Returns may be made in the form of a computer printout The printout should be in the same format and order as the questionnaire with each item identified by the identification number given against that line For example the Number for question 3 1 is 511, for 4 2 is 515, etc Space should be left on the printout for us to insert lines 501, 502, 505 and 504 at the top of Part A; lines 539 and 540 at the top of Part B; and lines 533 and 534 at the top of Part C If there are any difficulties please contact us

NOTES ON INDIVIDUAL QUESTIONS

1 DATE OF RETURN – QUESTION 1

Where possible give details for 31 December 1984 If these details are not readily available please give values for your annual accounting period ending after 30 June 1984 but before 1 July 1985 Please state the date to which the figures relate

2 GENERAL PARTICULARS – QUESTIONS 2, 6 AND 9

a For the purpose of this return

i the definition applied to a subsidiary is that given in Section 154 of the Companies Act 1948;

ii a branch means a permanent establishment as defined for the purpose of UK tax and UK double taxation agreements, being a fixed place of business overseas through which the business of the UK enterprise is wholly or partly carried on, and which is not a subsidiary Companies incorporated in the UK but operating wholly or mainly abroad should regard their permanent establishments abroad as overseas branches,

iii an associate company is an overseas company of which the UK reporting group or company owns part of the equity capital and which is an associated company of the UK reporting group or company as defined in the Statement of Standard Accounting Practice No 1,

iv United Kingdom covers Great Britain, Northern Ireland; overseas subsidiaries, branches and associates are those situated outside of Great Britain, Northern Ireland, Isle of Man, and the Channel Islands.

b Separate figures are required for each overseas branch of the UK reporting group or company, and for each overseas subsidiary or associate company in which the UK reporting group or company has a direct shareholding.

c If an overseas subsidiary company which is directly owned by the UK reporting group or company has subsidiaries of its own (called indirectly owned subsidiaries), please give consolidated figures for the directly owned company and all its subsidiaries, even if some of these indirectly owned subsidiaries are in a different country from the directly owned company If this is not possible you may give separate figures for each directly and indirectly owned company, as set out below. For an indirectly owned subsidiary state in question 2.1 the country in which that subsidiary is incorporated If you need additional forms please let us know.

If separate figures are given for directly and indirectly owned overseas subsidiaries

i In the separate column of figures you give for the indirectly owned subsidiaries enter in question 4 4 the value of the proportion of share capital, reserves and preference shares indirectly attributable to the UK reporting group or company The rest of the share capital, etc , of the indirectly owned subsidiaries is covered by voluntary question 4 5

ii In the separate figures for the directly owned subsidiary you may adopt one of the following methods to avoid double counting in the results

EITHER Enter in question 3 2 the value of the directly owned company's investment in the indirectly owned subsidiaries (This information is used to adjust for any double counting in the aggregate results for total overseas direct investment)

OR If you prefer, you may adjust the balance sheet by deducting the value of the directly owned company's investment in its subsidiaries from the directly owned company's share capital and reserves in questions 4 4 and 4 5 and leave question 3 2 blank

If an indirectly owned subsidiary has subsidiaries of its own for which separate figures are shown elsewhere in the return, treat the indirectly owned company's investment in its subsidiaries in the same way as for directly owned companies.

d If two or more overseas subsidiaries, branches or associates are engaged in the same type of business in the same country, a combined figure may be given. In such cases the percentage holding in question 2.4 should be calculated as the proportion of the share capital, reserves and preference shares of the combined subsidiaries that is directly and indirectly attributable to the UK reporting company or group. The percentage holding in question 9.3 should be calcul ated as the proportion you hold of the combined sum of the nominal value of the equity share capital of two or more associates

e. If you have an overseas subsidiary which you directly own but whose accounts are not consolidated in your group's consolidated accounts, please enter the net value of your investment in this unconsolidated subsidiary in Part C if you have insufficient details to complete Part A.

3. ASSETS – QUESTIONS 3 AND 7

a. Include goodwill, and overseas company's investment in its associated companies, and in its unconsolidated subsidiaries not included in question 3.2. Include intangible assets and any long-term loans owing to overseas concerns other than those owing by UK reporting group or company which should be included with current assets (question 3.3 or question 7.2). Deduct the total value of any government grants on these assets held as deferred credits or reserves.

b. Exclude loans and net current assets due to overseas company from its subsidiaries; these should be included in current assets and liabilities, as appropriate.

c. Include long and short-term loans owing by UK reporting group or company Include stocks, debtors, short-term loans, bank balances, deferred assets and cash.

4 LIABILITIES – QUESTIONS 4 AND 8

a. Include bank overdrafts and short-term loans, trade and other creditors, dividends and interest due, current taxation, short-term borrowing and any other current liabilities. Do not include deferred tax reserves.

b. Include under long-term liabilities any long-term or other loans not included in current liabilities at question 4 1. Include in question 4.3 or question 8 2 loans from other overseas members of the group and all long-term loans, debentures, etc , raised overseas by the overseas subsidiary or branch.

c. You are not required under the Statistics of Trade Act 1947 to provide the information requested in questions 4.5/6/7 and questions 8.3/4 but it would be bery helpful if you could give this information to complete the balance sheet for the overseas subsidiary or branch.

If you answer question 4.5 you do not have to answer question 2.4 since this can be calculated from the information you give in reply to questions 4 4 and 4 5

d. The net worth of the overseas branch consists of its fixed assets (net of depreciation) and its investments, plus its net current assets which are not due from or to the UK reporting group or company, less long-term loans other than from the UK reporting group or company less deferred tax and deferred credits of the branch and less any provision held by the branch. This should equal the unremitted profits and capital reserves of the branch, plus loans and net current account indebtedness due to the UK group or company.

PLEASE KEEP THESE NOTES FOR REFERENCE

A **compulsory** inquiry conducted by the Government Statistical Service
IN CONFIDENCE

DEPARTMENT OF TRADE AND INDUSTRY
Statistics Division 2
Room 252, Sanctuary Buildings
20 Great Smith Street
LONDON SW1P 3DB

Telephone 01-215 6235/6293

Date February 1987

Ref	Sub	YE	UKInd	Tp
1		86	21	

OT/OIL/OUT/86

← Please quote this reference number in any correspondence

Please correct any errors in name or address

NOTICE UNDER SECTION 1 OF THE STATISTICS OF TRADE ACT 1947

The Secretary of State for Trade and Industry hereby requires you by law to provide the information called for in this inquiry form with the exception of questions 5,6,9,10,11,12 and 13. These questions are voluntary and you are not required by law to answer them, but it would be helpful if you could, as all the information requested in this form is required to compile the balance of payments, for the appreciation of economic trends and for the provision of a statistical service for government and industry.

Please complete this form and return it to the Department of Trade and Industry by Friday 22 May 1987

N.B. The information given by you under the Act will be treated as confidential in strict accordance with the Act and subject to the further restriction that information about individual businesses will be used and disclosed under Ministerial direction to other government departments for statistical purposes only, except that the names and addresses of individual businesses and their industrial classification may, if a Minister so directs, be made available to Ministers and officials of government departments who need them for carrying out their functions

ANNUAL INQUIRY INTO TRANSACTIONS OF UNITED KINGDOM OIL COMPANIES WITH THEIR OVERSEAS SUBSIDIARIES, ASSOCIATES AND BRANCHES IN 1986

Dear Contributor

We are conducting this annual inquiry to obtain up-to-date statistical information about various overseas transactions and earnings of United Kingdom oil companies for use in compiling our country's balance of payments statistics. The information requested is the type needed by any government when deciding its economic policy as well as its commercial policy in relation to other countries. This inquiry is the only source of such information and without it the accuracy of the UK balance of payments statistics will be greatly reduced.

The form should be returned even where there is little or nothing to report against some of the questions. I should like to emphasise that if you do not have precise figures readily available we are happy to accept your best estimates. We are prepared to accept your return in the form of a computer printout if you prefer - please see introduction on the enclosed notes.

I would be grateful if you could return the completed form to us as soon as possible. In any case it should be returned by Friday 22 May 1987.

There are notes enclosed to help you complete your return. If you have any difficulties or require more information about the inquiry my staff will be pleased to help you. The telephone extensions of those dealing with this inquiry are shown above.

Yours faithfully

P H Richardson.

P. H. RICHARDSON
Chief Statistician

SC408A

**IMPORTANT Please refer to the enclosed notes before completing this form.
If you do not have precise figures readily available give the best estimates you can.
Please show all values in sterling rounded in £ million to one decimal place.
Use the same rates of conversion as in your own accounts.**

Please give figures for the year ending 31 December 1986.

PART A - OVERSEAS SUBSIDIARY AND ASSOCIATE COMPANIES

1. **GENERAL PARTICULARS** (see note 1)

1.1 Country in which overseas subsidiary or associate is incorporated........

1.2 Country of ownership of overseas subsidiary or associate........
If company is directly owned by UK reporting group or company write "UK" in the box; if indirectly owned through another overseas company give country of that company directly owning the subsidiary or associate.

2. **AMOUNTS RECEIVED FROM AND PAID TO OVERSEAS SUBSIDIARIES AND ASSOCIATES** by UK reporting group or company

2.1 **Interest received** from overseas company net of any withholding tax paid overseas, less interest paid to that company net of any UK withholding tax you deducted in 1986 (see note 2 1)........

2.2 **Dividends received** from overseas company by UK reporting group or company less any withholding tax deducted overseas, plus any refunds of this withholding tax on current and previous years' dividends received in year (see note 2.2)........

3. **PURCHASES AND SALES OF THE OVERSEAS COMPANIES' SHARES, DEBENTURES, AND QUOTED LOAN STOCK** (including preference shares) by UK reporting group or company. (Please give cash values not nominal values.) (see note 3).

3.1 Purchases (including new subscriptions)........

3.2 Sales (including proceeds from redemptions)........

4. **NET TOTAL OF INTER-COMPANY CURRENT ACCOUNT BALANCES, LONG-TERM LOANS** (other than debentures and quoted loan stock) and **SHORT TERM LOANS** outstanding between the overseas company and the UK reporting group or company (see note 4).

4.1 At 31.12.86. Net amount due from UK reporting group or company........
or
Net amount due to UK reporting group or company........

4.2 At 31.12.85. Net amount due from UK reporting group or company........
or
Net amount due to UK reporting group or company........

5. **UK REPORTING GROUP'S OR COMPANY'S SHARE OF OVERSEAS SUBSIDIARY'S OR ASSOCIATE'S PROFIT OR LOSS** (whether remitted or not) after allowing for provisions for depreciation, accrued tax and extraordinary items (see note 5).
This question is voluntary.

5.1 Profit........
or
5.2 Loss........

6. **NET GAIN OR LOSS RESULTING FROM FOREIGN EXCHANGE RATE MOVEMENTS** made by the UK reporting group or company on the balances reported in question 4 above (see note 6).
This question is voluntary.

6.1 Gain........
or
6.2 Loss........

Please also complete Part B overleaf and Part C attached

OT/OIL/OUT/86

For official use

	Ref	Sub		Form
5			86	

	1	2	3	4	5	6	7	8
61								
62								
60								

....								
....								

	TOTAL £ million	£ million	£ million	£ million	£ million	£ million	£ million	£ million	£ million
.... 65									
.... 66									
.... 73									
.... 74									
.... 76									
.... 77									
.... 78									
.... 79									
.... 63									
.... 64									
.... 471									
.... 472									

PLEASE COMPLETE THIS FORM AND RETURN IT BY 22 MAY 1987

For official use

Ref		Sub	Form	
6			86	

	1	2	3
21			
22			

PART B – OVERSEAS BRANCHES

7. GENERAL PARTICULARS (see note 1)

Country in which overseas branch is situated ....

8. NET ASSETS OF OVERSEAS BRANCH ATTRIBUTABLE TO UK REPORTING GROUP OR COMPANY at end of year as shown in the accounts of the UK reporting group or company. (Include loans and current account balances of branch due to or from other UK companies in your group) (see note 8).

			TOTAL £ million	£ million	£ million	£ million
8.1 At 31.12.86	Net amount due from UK reporting group or company	26				
or	Net amount due to UK reporting group or company	27				
8.2 At 31 12.85.	Net amount due from UK reporting group or company	28				
or	Net amount due to UK reporting group or company	29				

9. TOTAL PROFIT OR LOSS OF OVERSEAS BRANCH ATTRIBUTABLE TO UK REPORTING GROUP OR COMPANY (whether remitted or not) after allowing for provisions for depreciation, accrued overseas tax and extraordinary items. (Add into profits any interest paid by branch to reporting company or other UK companies in group less any interest paid to overseas branch) (see note 9).
This question is voluntary.

9.1 Profit	23				
or 9.2 Loss	24				

10. NET GAIN OR LOSS RESULTING FROM FOREIGN EXCHANGE RATE MOVEMENTS made by UK reporting group or company on the net assets of the branch and reported in question 8 above (see note 6).
This question is voluntary.

10.1 Gain	473				
or 10.2 Loss	474				

Please also complete Part C attached

A **compulsory** inquiry conducted by the Government Statistical Service
IN CONFIDENCE

DEPARTMENT OF TRADE AND INDUSTRY
Statistics Division 2
Room 252, Sanctuary Buildings
20 Great Smith Street
LONDON SW1P 3DB

Telephone 01-215 6235/6293
Date February 1987

Ref	Sub	YE	UKInd	Tp
1		86	21	

CA/OIL/OUT/86

← Please quote this reference number in any correspondence

Please correct any errors in name or address

NOTICE UNDER SECTION 1 OF THE STATISTICS OF TRADE ACT 1947

The Secretary of State for Trade and Industry hereby requires you by law to provide the information called for in this inquiry form with the exception of questions 3.5/6/7 and 7.3/4. These questions are voluntary and you are not required by law to answer them, but it would be helpful if you could, as all the information requested in this form is required to compile the balance of payments, for the appreciation of economic trends and for the provision of a statistical service for government and industry.

Please complete this form and return it to the Department of Trade and Industry by Friday 22 May 1987

N.B. The information given by you under the Act will be treated as confidential in strict accordance with the Act and subject to the further restriction that information about individual businesses will be used and disclosed under Ministerial direction to other government departments for statistical purposes only, except that the names and addresses of individual businesses and their industrial classification may, if a Minister so directs, be made available to Ministers and officials of government departments who need them for carrying out their functions

ASSETS AND LIABILITIES AT END 1986
OF OVERSEAS SUBSIDIARIES, BRANCHES AND ASSOCIATES OF UK OIL COMPANIES

Dear Contributor

We are conducting this annual inquiry to obtain information on the value of United Kingdom companies' investment in their overseas subsidiaries, branches and associates. It is the only source of such data and is used in compiling the UK balance of payments statistics and in deciding government policy in relation to other countries.

Please note that questions 3.5/6/7 and 7.3/4 are voluntary and you are not required by law to answer them. However it would be very helpful if you could give this information as it will complete the balance sheets for the overseas subsidiaries or branches. Any information given voluntarily will be treated as confidential. It will not be disclosed outside this office and will be used only for compiling totals which will not disclose the figures for individual contributors.

If the total value of your overseas investments in all countries is between minus £500,000 and plus £500,000 just tick the box indicated, sign and return the form to us. You do not have to answer any other questions.

The form should be returned even where there is little or nothing to report against some of the questions. I should like to emphasise that if you do not have precise figures readily available we are happy to accept your best estimates. We are prepared to accept your return in the form of a computer printout if you prefer – please see introduction on the enclosed notes.

I would be grateful if you could return the completed form to us as soon as possible. In any case it should be returned by Friday 22 May 1987.

There are notes enclosed to help you complete your return. If you have any difficulties or require more information about the inquiry my staff will be pleased to help you. The telephone extensions of those dealing with this inquiry are shown above.

Yours faithfully

P H Richardson.

P. H. RICHARDSON
Chief Statistician

SC404A

IMPORTANT Please refer to the enclosed notes before completing this form.
If you do not have precise figures readily available give the best estimates you can.
Please show all values in sterling rounded in £ million to one decimal place
(e.g. £53,440,000 should be entered as 53.4). Use the same rates of conversion as in your own accounts.
Please give details for 31 December 1986.

If you and your UK subsidiaries have no overseas subsidiaries, branches or associates tick this box ☐ sign and return the form. You do not have to answer any other questions.

If the total value of your investments in all countries combined (questions 3.2 + 3.4 + 4.2 − 4.1 + 7.4 + 9) is between minus £500,000 and plus £500,000 tick this box ☐ sign and return the form. You do not have to answer any other questions.

PART A – OVERSEAS SUBSIDIARY COMPANIES

1. GENERAL PARTICULARS (see note 1)

1.1 Country in which overseas subsidiary is incorporated ....

1.2 Main activity of overseas subsidiary ....

1.3 Country of ownership of overseas subsidiary (see notes 1a and 1b) ....

If company is directly owned by UK reporting group or company write "UK" in box; if indirectly owned through another overseas subsidiary give country of that subsidiary directly owning the company.

1.4 Percentage of overseas subsidiary's share capital and reserves, including preference shares, attributable to UK reporting group or company (see note 1c). (You do not have to answer this question if you answer question 3.5 below.) ....

2. ASSETS

2.1 Fixed assets after provision for depreciation plus investments in securities (see note 2a) ....

2.2 Value of investment by overseas subsidiary in its subsidiaries for which separate figures are given elsewhere in this return (see note 2) (see note 1b for alternative treatment of these subsidiaries) ....

2.3 Current assets (see note 2c) ....

2.4 **TOTAL ASSETS** (sum of 2.1 to 2.3) (Should equal total liabilities at 3.8) ....

3. LIABILITIES

3.1 Current liabilities (see note 3a) ....

3.2 Long-term liabilities owing to UK reporting group or company (see note 3b) ....

3.3 Long-term liabilities other than those owing to UK reporting group or company (see note 3b) ....

3.4 Amount of share capital, reserves, and preference shares of overseas subsidiary attributable to the UK reporting group or company ....

3.5 Amount of share capital, reserves including preference shares, of overseas subsidiary attributable to other shareholders (see note 3c) (This question is voluntary) ....

3.6 Minority shareholders' interests in any of the overseas company's subsidiaries consolidated in the above figures (see note 3c) (This question is voluntary) ....

3.7 Provisions, deferred tax reserves, and any other liabilities (see note 3c) (This question is voluntary) ....

3.8 **TOTAL LIABILITIES** (sum of 3.1 to 3.7) (Should equal total assets at 2.4) ....

4. INTER-COMPANY ACCOUNT

Inter-company account balances with UK reporting group or company included in

4.1 Current assets at 2.3 above ....

4.2 Current liabilities at 3.1 above ....

For official use **CA/OIL/OUT/86**

Ref		Sub	Form
2		86	

	1	2	3	4	5	6	7	8
501								
502								
505								

504	%	%	%	%	%	%	%	%

	TOTAL £ million	£ million	£ million	£ million	£ million	£ million	£ million	£ million	£ million
511									
512									
513									
506									
514									
515									
516									
535									
520									
510									
517									
507									
518									
519									

For official use

Ref		Sub	Form
3		86	

	1	2	3
539			
540			

PART B – OVERSEAS BRANCHES

5. GENERAL PARTICULARS (see note 1a)

5.1 Country in which overseas branch is situated ....

5.2 Main activity of overseas branch ....

		TOTAL £ million	£ million	£ million	£ million
6. ASSETS					
6.1 Fixed assets after provision for depreciation plus investments in securities (see note 2a)	521				
6.2 Current assets (see note 2c)	523				
6.3 **TOTAL ASSETS** (6.1 plus 6.2)	542				
7. LIABILITIES					
7.1 Current liabilities excluding amounts due to UK reporting group or company (see note 3a)	524				
7.2 Long-term liabilities other than to UK reporting group or company (see note 3b)	525				
7.3 Provisions, deferred tax reserves, and any other liabilities (see note 3c) (This question is voluntary)	526				
7.4 Net worth of overseas branch attributable to UK reporting group or company (see notes 3c and 3d) (This question is voluntary)	536				
7.5 **TOTAL LIABILITIES** (sum of 7.1 to 7.4) (Should equal total assets at 6.3 above)	543				

For official use

Ref		Sub	Form
4		86	

	1	2	3
533			
534			

PART C – OVERSEAS ASSOCIATES

Include only overseas associates directly owned by UK reporting group or company

8. GENERAL PARTICULARS (see notes 1a and 1d)

8.1 Country in which overseas associate company is situated ....

8.2 Main activity of overseas associate ....

8.3 Percentage of issued share capital owned by the UK reporting group or company (see note 1c) .... 541 | % | % | %

		TOTAL £ million	£ million	£ million	£ million
9. VALUE OF INVESTMENT Value of investment in overseas associate as valued under SSAP1 in the accounts of the UK reporting group or company plus loan and net current account balances due from the overseas associate	527				

Name of person to be contacted should this be necessary ____________________
BLOCK CAPITALS PLEASE

Position in company ____________________ Signature ____________________

Telephone No/Ext ____________________ Telex ____________________ Date ____________________

SC404D

For official use

Ref		Sub	Form	
11			86	

OT/OIL/OUT/86

PART C – OTHER OVERSEAS TRANSACTIONS ATTRIBUTABLE TO UNITED KINGDOM OIL COMPANIES

11. **TOTAL NET TRADE CREDIT RECEIVED BY UK REPORTING GROUP OR COMPANY** in respect of purchases of all goods and services (including credit received on goods purchased and held outside the UK) from non-UK residents. Include credit received from non-UK financial institutions as well as non-group overseas suppliers, and subtract any advance and progress payments made (see note 11). This question is voluntary

			£ million
11.1	Net credit received outstanding at 31.12.86	423	
11.2	Net credit received outstanding at 31.12.85	424	

12. **TOTAL NET TRADE CREDIT EXTENDED BY UK REPORTING GROUP OR COMPANY** in respect of sales of all goods and services (including credit on goods bought and sold outside the UK) to non-UK residents. Include credit extended to all non-group overseas customers and subtract all advance and progress payments received (see note 12). This question is voluntary.

12.1	Net credit extended outstanding at 31.12.86	433	
12.2	Net credit extended outstanding at 31 12.85	434	

13. **CHANGES IN STOCKS OF OIL AND OIL PRODUCTS HELD OUTSIDE THE UK BY UK REPORTING GROUP OR COMPANY** (see note 13). This question is voluntary

13.1	Increase during 1986	443	
	or		
13.2	Decrease during 1986	444	

14. **OVERSEAS ROYALTIES AND SIMILAR TRANSACTIONS** on revenue and capital account. Include overseas transactions with non-group companies as well as those with your subsidiaries, associates and branches overseas. Exclude receipts for goods imported into or exported from the UK (see note 14).

14.1	Total amount received from overseas less any tax paid overseas in 1986	491	
14.2	Total amount paid overseas less any UK tax you deducted in 1986	492	

15. **OTHER OVERSEAS TRANSACTIONS** with non-group companies as well as those with your subsidiaries, associates and branches overseas. Include merchanting of oil and oil products, management and head office expenses, accountants' fees, directors' fees, agent's commission, advertising, insurance premiums and claims, etc. Exclude UK imports and exports, their shipping and freight costs, cinematograph film and TV material, and interest and dividend payments (see note 14).

15.1	Total amount received from overseas less any tax paid overseas in 1986	493	
15.2	Total amount paid overseas less any UK tax you deducted in 1986	494	

Please return this page with your main form (Parts A and B)

Name of person to be contacted should this be necessary ----------------------------
BLOCK CAPITALS PLEASE

Position in company ------------------------------ Signature ----------------

Telephone No/Ext -------------------------------- Telex --------- Date ------

SC408E

OT/OIL/OUT/86N

ANNUAL INQUIRY INTO TRANSACTIONS OF UNITED KINGDOM OIL COMPANIES WITH THEIR OVERSEAS SUBSIDIARIES, ASSOCIATES AND BRANCHES IN 1986

NOTES ON COMPLETING FORM OT/OIL/OUT/86

INTRODUCTION

UK parent companies should complete this return on behalf of all UK companies within their group.

Returns may be made in the form of a computer printout. The printout should be in the same format and order as the questionnaire, with each item identified by the identification number given against that line. For example, the Number for question 2.1 is 65, for 2.2 is 66, etc. Space should be left on the printout for us to insert lines 61,62 and 60 at the top of Part A, and lines 21 and 22 at the top of Part B. If there are any difficulties please contact us.

DEFINITIONS

For the purpose of this return:

a. the definition applied to a **subsidiary** is that given in Section 736 of the Companies Act 1985; (this is the same definition as that previously given in Section 154 of the Companies Act 1948.)

b. an **associate company** is an overseas company of which the UK reporting group or company directly owns part of the equity capital and which is an associated company of the UK reporting group or company as defined in the Statement of Standard Accounting Practice No 1 (SSAP1).

c. a **branch** means a permanent establishment as defined for the purpose of the UK tax and UK double taxation agreements, being a fixed place of business overseas through which the business of the UK enterprise is wholly or partly carried on, and which is not a subsidiary. Include as branches joint ventures carried out overseas with UK or overseas companies if these are not conducted through an overseas subsidiary or associate company. Companies incorporated in the UK but operating wholly or mainly abroad should regard their permanent establishments abroad as overseas branches.
A UK incorporated company with its head office overseas should regard its overseas head office as an overseas branch and its UK registered office as its UK head office and complete Part B of this form accordingly.

d. **overseas** subsidiaries, associates and branches are those situated outside of the United Kingdom. The Republic of Ireland is regarded as an overseas country and any subsidiary, associate or branch in that country should be included in this return.

e. **United Kingdom** covers Great Britain and Northern Ireland, the Isle of Man, the Channel Islands, and that part of the Continental Shelf which is defined by International Convention as belonging to the UK.

VALUATION

Show values in sterling as entered in the UK reporting group's or company's accounts in £million rounded to one decimal place. If precise figures are not readily available please give the best estimates you can. Where conversion from an overseas currency is involved, please use the same rate of exchange as in your company's own accounts.

NOTES ON INDIVIDUAL QUESTIONS

1. GENERAL PARTICULARS – QUESTIONS 1 AND 7

If you have subsidiary companies incorporated in the UK please complete this questionnaire on behalf of all UK companies within your group. If this is not possible please complete this return in respect of your company and arrange for your UK subsidiaries to make separate returns.
Separate figures are required for each country in which you have a subsidiary, associate or branch. If you have more than one subsidiary, associate or branch in a country you can either give a single combined figure for that country or, if it is easier for you, separate figures for each subsidiary, associate or branch.
If you have an overseas subsidiary which is held indirectly through another overseas subsidiary you may either give consolidated figures for each directly owned company covering that company and all its subsidiaries and associates even if some of these are in different countries, or you may give separate figures for each directly and indirectly owned subsidiary.
Include in your return any subsidiaries you have established overseas to raise funds on the international capital market.
Please let us know if you need additional forms.

2. INTEREST AND DIVIDENDS

2.1 Please state net interest received and not interest "accrued". If amounts paid exceed those received enter figures as minus (–).

2.2 State dividends received in the year including any declared in earlier years and paid this year. Do not include any dividends declared but not received in the year in either question 2.2 or in balances outstanding in question 4.
Include dividends on preference shares and any tax refunds received directly from overseas Revenue Authorities in 1986.

3. PURCHASES AND SALES OF SHARES ETC

Include transactions in quoted and unquoted shares and debentures of the overseas company plus any in its loan stock quoted on a stock exchange in the UK or overseas. Also include purchases made by issuing shares of your UK group of companies in exchange for the equity of an overseas company (valuing this, if possible, at the market value on the day of acquisition of the shares issued), as well as acquisitions made in return for equipment, manufacturing rights, securities, etc, supplied by the UK company to overseas companies. In addition include any capital injections which are not for repayment and on which no interest, etc, is payable. Bonus issues of shares arising from the capitalisation of reserves should be omitted but any acquisition of shares or debentures or quoted loan stock arising from the capitalisation of inter-company current account balances or loans should be included in question 3.1.

4. INTER-COMPANY ACCOUNT BALANCES

Include unquoted loan stock. Exclude any dividends declared but not received in the year and any bonus issues of shares arising from the capitalisation of reserves. Any acquisition of shares or debentures or quoted loan stock arising from the capitalisation of inter-company current account balances or loans should be included in question 3.1 with a corresponding fall shown in the amount due to you in question 4.

5. SHARE OF PROFIT OR LOSS

The share of the profits of overseas subsidiary and associate companies should be determined in accordance with the UK reporting group or company's proportionate holding of the subsidiary or associate company's equity share capital. Please contact us if you have any difficulties in determining this proportion (e.g various classes of voting shares). Make allowances for depreciation, accrued overseas tax and extraordinary items. Deduct withholding taxes paid overseas on dividends net of any refunds of this tax received in the year, including any refunds received directly from overseas Revenue Authorities.

6. NET GAIN OR LOSS RESULTING FROM FOREIGN EXCHANGE RATE MOVEMENTS - QUESTIONS 6 AND 10

If these figures are not available, please estimate, or if no estimate is possible enter "N/A". If zero, please enter "NIL".

8. NET ASSETS OF OVERSEAS BRANCH ATTRIBUTABLE TO UK REPORTING GROUP OR COMPANY

State the net assets attributable to the UK reporting group or company . These consist of its fixed assets (net of depreciation) and its investments, plus its net current assets with concerns other than the UK reporting group or company, less long-term loans and other liabilities (accrued tax, deferred credits etc.) held by the branch other than those due to the UK reporting group or company This should equal the unremitted profits and capital reserves of the branch plus loans and net current account indebtedness due to the UK reporting group or company.

9. TOTAL PROFIT OR LOSS OF OVERSEAS BRANCH ATTRIBUTABLE TO UK REPORTING GROUP OR COMPANY

State the total profit or loss of overseas branch attributable to UK reporting group or company whether remitted or not Make allowances for depreciation, accrued overseas tax and extraordinary items. Also deduct amounts payable, net of any refunds, of overseas withholding tax. Add into profits any interest paid by branch to reporting company or other UK companies in group, less any interest paid to branch by reporting company or other UK companies in group.

11. TOTAL NET TRADE CREDIT RECEIVED BY UK REPORTING GROUP OR COMPANY

Include trade credit received by the UK reporting group or company in respect of purchases from non-UK residents of goods and services. Exclude trade credit received from overseas subsidiaries and associates - such credit should be reflected in inter-company balances in question 4; exclude trade credit received from overseas branches - such credit should be reflected in the net assets of branches in question 8.

Example: If a tanker of crude oil is purchased by the UK company from a non-UK resident outside the group in the second week of December under a contract requiring payment in 30 days, then the credit received on the purchase will still be outstanding at the end of the year.

12. TOTAL NET TRADE CREDIT EXTENDED BY UK REPORTING GROUP OR COMPANY

Include trade credit extended by the UK reporting group or company in respect of sales to non-UK residents of goods and services. Exclude trade credit extended to overseas subsidiaries and associates - such credit should be reflected in inter-company balances in question 4; exclude trade credit extended to overseas branches - such credit should be reflected in the net assets of branches in question 8.

Example: If a tanker of crude oil is sold by the UK company to a non-UK resident outside the group in the second week of December under a contract requiring payment in 30 days, then the credit extended on the sale will still be outstanding at the end of the year.

13. CHANGES IN STOCKS HELD OVERSEAS

Exclude stocks of goods held through overseas subsidiaries and associates.

14. OVERSEAS ROYALTIES AND OTHER TRANSACTIONS - QUESTIONS 14 AND 15

For royalties in question 14:-

Include

(i) royalties on printed matter, sound recordings, performing rights, extraction/drilling processes, patents, licences, trade marks, designs, copyright, mining and quarrying mineral rights (but not oil royalties), manufacturing rights, research and development, technical know-how and assistance.

(ii) lump-sum royalty settlements made in year.

(iii) amounts payable to overseas agents for charges (less discounts) for services rendered in connection with royalties and similar transactions listed under (i) and (ii) above.

(iv) total value of royalties, etc. which you collect from, or pay to, overseas residents even though you may be acting as an agent.

Exclude

(i) oil royalties

(ii) royalties, etc. included in the invoiced price of goods you bought or sold as these are covered by other statistics;

(iii) royalties, rentals, etc. in respect of cinematograph film and television programme material as these are covered by another inquiry.

If it is not possible to split a payment or receipt between royalties (question 14) and other transactions (question 15) (e.g where amounts include both royalties and other services such as advertising, market research, use of distribution and marketing facilities, etc) please enter the total amount in question 14 if royalties account for half or more, or in question 15 if royalties account for less than half.

PLEASE KEEP THESE NOTES FOR REFERENCE SC408G

CA/OIL/OUT/86N

OVERSEAS ASSETS AND LIABILITIES AT END 1986
OF OVERSEAS SUBSIDIARIES, BRANCHES AND ASSOCIATES OF UK OIL COMPANIES

NOTES ON COMPLETING FORM CA/OIL/OUT/86

INTRODUCTION

UK parent companies should complete this return on behalf of all UK companies within their group, unless alternative arrangements were made for the previous inquiry. Those groups which would like their subsidiary UK companies to make separate returns should ensure that the parent company makes its own return (where applicable) and arrange for their subsidiary UK companies to make separate returns. Please advise us accordingly and we will supply the necessary additional forms.

Returns may be made in the form of a computer printout. The printout should be in the same format and order as the questionnaire with each item identified by the identification number given against that line. For example the Number for question 2.1 is 511, for 3.2 is 515, etc. Space should be left on the printout for us to insert 501, 502, 505 and 504 at the top of Part A; lines 539 and 540 at the top of Part B; and lines 533 and 534 at the top of Part C If there are any difficulties please contact us.

DEFINITIONS

For the purpose of this return:

a. the definition applied to a **subsidiary** is that given in Section 736 of the Companies Act 1985; (this is the same definition as that previously given in section 154 of the Companies Act 1948).

b. a **branch** means a permanent establishment as defined for the purpose of UK tax and UK double taxation agreements, being a fixed place of business overseas through which the business of the UK enterprise is wholly or partly carried on, and which is not a subsidiary. Companies incorporated in the UK but operating wholly or mainly abroad should regard their permanent establishments abroad as overseas branches

c. an **associate company** is an overseas company of which the UK reporting group or company directly owns part of the equity capital and which is an associated company of the UK reporting group or company as defined in the Statement of Standard Accounting Practice No 1 (SSAP1).

d. **United Kingdom** covers Great Britain, Northern Ireland and UK Continental Shelf; overseas subsidiaries, branches and associates are those situated outside of Great Britain, Northern Ireland, Isle of Man, and the Channel Islands.

VALUATION

Show values in sterling as entered in the UK reporting group's or company's accounts in £ million rounded to one decimal place. If precise figures are not readily available please give the best estimates you can. Where conversion from an overseas currency is involved, please use the same rate of exchange as in your company's own accounts.

NOTES ON INDIVIDUAL QUESTIONS

1. GENERAL PARTICULARS – QUESTIONS 1,5 AND 8

a. Separate figures are required for each overseas branch of the UK reporting group or company, and for each overseas subsidiary or associate company in which the UK reporting group or company has a direct shareholding.

b. If an overseas subsidiary company which is directly owned by the UK reporting group or company has subsidiaries of its own (called indirectly owned subsidiaries), please give consolidated figures for the directly owned company and all its subsidiaries, even if some of these indirectly owned subsidiaries are in a different country from the directly owned company. If this is not possible you may give separate figures for each directly and indirectly owned company, as set out below. For an indirectly owned subsidiary state in question 1.1 the country in which that subsidiary is incorporated. If you need additional forms please let us know.

If separate figures are given for directly and indirectly owned overseas subsidiaries:

i. In the separate column of figures you give for the indirectly owned subsidiaries enter in question 3.4 the value of the proportion of share capital, reserves and preference shares indirectly attributable to the UK reporting group or company. The rest of the share capital, etc., of the indirectly owned subsidiaries is covered by voluntary question 3.5.

ii. In the separate figures for the directly owned subsidiary you may adopt one of the following methods to avoid double counting in the results.

EITHER: Enter in question 2.2 the value of the directly owned company's investment in the indirectly owned subsidiaries. (This information is used to adjust for any double counting in the aggregate results for total overseas direct investment.)

OR: If you prefer, you may adjust the balance sheet by deducting the value of the directly owned company's investment in its subsidiaries from the directly owned company's share capital and reserves in questions 3.4 and 3.5 and leave question 2.2 blank.

If an indirectly owned subsidiary has subsidiaries of its own for which separate figures are shown elsewhere in the return, treat the indirectly owned company's investment in its subsidiaries in the same way as for directly owned companies.

c. If two or more overseas subsidiaries, branches or associates are engaged in the same type of business in the same country, a combined figure may be given. In such cases the percentage holding in question 1.4 should be calculated as the proportion of the share capital, reserves and preference shares of the combined subsidiaries which is directly and indirectly attributable to the UK reporting company or group. The percentage holding in question 8.3 should be calculated as the proportion you hold of the combined sum of the nominal value of the equity share capital of two or more associates.

d. If you have an overseas subsidiary which you directly own but whose accounts are not consolidated in your group's consolidated accounts, please enter the net value of your investment in this unconsolidated subsidiary in Part C if you have insufficient details to complete Part A.

SC404E

2. ASSETS - QUESTIONS 2 AND 6

a Include in question 2.1 overseas company's investment in its associated companies, and in its unconsolidated subsidiaries not included in question 2.2 Include in question 2.1 and 6.1 goodwill, other intangible assets and any long-term loans owing to the directly owned overseas subsidiaries or branches other than those owing by UK reporting group or company which should be included with current assets (question 2.3 or question 6.2). Deduct the total value of any government grants on these assets held as deferred credits or reserves.

b Exclude loans and net current assets due to overseas company from its subsidiaries; these should be included in current assets and liabilities, as appropriate.

c. Include long and short-term loans owing by UK reporting group or company. Include stocks, debtors, short-term loans, bank balances, deferred assets and cash

3. LIABILITIES - QUESTIONS 3 AND 7

a Include bank overdrafts and short-term loans, trade and other creditors, dividends and interest due, current taxation short-term borrowing and any other current liabilities. Do not include deferred tax reserves.

b Include under long-term liabilities any long-term or other loans not included in current liabilities at question 3.1 or 7.1 as appropriate Include in question 3.3 or question 7.2 loans from other overseas members of the group and all long-term loans, debentures, etc., raised overseas by the overseas subsidiary or branch

c You are not required under the Statistics of Trade Act 1947 to provide the information requested in questions 3.5/6/7 and 7.3/4 but it would be very helpful if you could give this information to complete the balance sheet for the overseas subsidiary or branch

If you answer question 3.5 you do not have to answer question 1.4 since this can be calculated from the information you give in reply to questions 3.4 and 3.5.

d The net worth of the overseas branch consists of its fixed assets (net of depreciation) and its investments, plus its net current assets which are not due from or to the UK reporting group or company, less long-term loans other than from the UK reporting group or company, less deferred tax and deferred credits of the branch and less any provision held by the the branch This should equal the unremitted profits and capital reserves of the branch, plus loans and net current account indebtedness due to the UK group or company.

PLEASE KEEP THESE NOTES FOR REFERENCE

SC404F

A voluntary inquiry conducted by
the Government Statistical Service
IN CONFIDENCE

DEPARTMENT OF TRADE AND INDUSTRY
Statistics Division 2
Room 252, Sanctuary Buildings
20 Great Smith Street
LONDON SW1P 3DB

Telephone. 01-215 6235/6293

Date February 1987

OT/OIL/IN/86

← Please quote this reference number when writing to us

ANNUAL INQUIRY INTO OVERSEAS TRANSACTIONS OF OIL COMPANIES OPERATING ONSHORE IN THE UK IN 1986

Dear Contributor

The purpose of this inquiry is to obtain up-to-date statistical information about various transactions between foreign owned oil companies operating onshore in the UK and their parent company overseas and overseas fellow subsidiaries and branches for use in compiling our country's balance of payments statistics. Similar transactions relating to interests in the UK sector of the Continental Shelf should be reported on a separate inquiry. The information is needed by government in deciding its economic policy as well as its commercial policy in relation to other countries. This inquiry is the only source of such information and without it the accuracy of the UK balance of payments statistics will be greatly reduced.

The current form is enclosed and I would be most grateful if you would complete it and return it to me preferably not later than 22 May this year.

All information supplied will be treated as confidential. Only those persons directly employed on collecting, processing and interpreting the information will have access to the particulars of individual returns, which will not be disclosed outside this small group of officials.

If you have any difficulties or would like any further information about this particular inquiry, my staff will be pleased to help you. The telephone extensions of those dealing with this inquiry are shown above.

Yours faithfully

P H Richardson.

P H RICHARDSON
Chief Statistician

SC406A

A voluntary inquiry conducted by the Government Statistical Service

IN CONFIDENCE

DEPARTMENT OF TRADE AND INDUSTRY
Statistics Division 2
Room 252, Sanctuary Buildings
20 Great Smith Street
LONDON SW1P 3DB

Telephone: 01-215 6235/6293

Date February 1987

Ref		Sub		YE	UKInd	Tp
1			86		21	

OT/OIL/IN/86

← Please quote this reference number when writing to us

For official use

	Ref	Sub	Form
10		86	

	1	2
161		

Please correct any errors in name or address

ANNUAL INQUIRY INTO OVERSEAS TRANSACTIONS OF OIL COMPANIES OPERATING ONSHORE IN THE UK IN 1986

IMPORTANT **Please refer to the enclosed notes before completing this form. If you do not have precise figures readily available give the best estimates you can. Please show all values in sterling rounded in £ million to one decimal place (e.g. £53,440,000 should be entered as 53.4). Use the same rates of conversion as in your own accounts.**

DATE OF RETURN : Please give details for 31 December 1986

1. COUNTRY IN WHICH OVERSEAS PARENT OR ASSOCIATE SHAREHOLDER IS INCORPORATED OR REGISTERED (see note 1) ......

2. OVERSEAS DIRECT INVESTMENT IN AND EARNINGS OF UK COMPANY

Item		Box	£ million	£ million
2.1 Overseas company's share of the UK reporting company's consolidated profit or loss (whether remitted or not) after allowing for provision for depreciation, accrued tax and extraordinary items (see note 2.1)	a. Profit	162		
	b. Loss	163		
2.2 Interest paid to, less interest received from, overseas parent company and overseas fellow subsidiaries and branches (see note 2.2)		164		
2.3 Dividends paid to overseas parent company (see note 2.3)		165		
2.4 UK tax credits paid directly by you to overseas parent company in respect of dividends (see note 2.4)		166		
2.5 Purchases and sales by the overseas parent company of the reporting company's shares (including preference shares) debentures and listed loan stock (see note 2.5)	a Purchases	173		
	b. Sales	174		
2.6 Net total of inter-company current account balances, long-term loans (other than debentures and listed loan stock) and short-term loans outstanding between UK reporting company and the overseas company, or overseas fellow subsidiaries, branches and associates (see note 2.6)	a Net amount due from UK company at end of current year ...	176		
	OR			
	b. Net amount due to UK company at end of current year ...	177		
	AND			
	c. Net amount due from UK company at end of previous year	178		
	OR			
	d Net amount due to UK company at end of previous year	179		

SC406B

PLEASE RETURN THIS FORM BY 22 MAY 1987

			£ million	£ million
2.7 Net gain or loss resulting from exchange rate movements made by the reporting company on the balances reported in question 2.6 (see note 2.7)	a. Gain	168		
	OR b. Loss	169		
2.8 Total corporation tax accrued during year (see note 2.8)		414		

3. OTHER OVERSEAS PAYMENTS AND RECEIPTS (see note 3)

3.1 Royalties and similar transactions on revenue and capital account. Include overseas transactions with non-group companies as well as those with group members (see note 3.1)	a. Total amount received from overseas less any tax paid overseas	203	
	b. Total amounts paid overseas less any UK tax you deducted	204	
3.2 All other transactions not elsewhere specified. Include overseas transactions with non-group companies as well as those with group members (see note 3.2)	a. Total amounts received from overseas less any tax paid overseas	213	
	b. Total amounts paid overseas less any UK tax you deducted	214	
3.3 Total net trade credit received by UK reporting group or company in respect of purchases of all goods and services (including credit received on goods purchased and held outside the UK) from non-UK residents (see note 3.3)	a. Net credit received outstanding at end of current year	421	
	b. Net credit received outstanding at end of previous year	422	
3.4 Total net trade credit extended by UK reporting group or company to non-UK residents in respect of sales of all goods and services (including credit extended on goods bought and sold outside the UK) (see note 3.4)	a. Net credit extended outstanding at end of current year	431	
	b. Net credit extended outstanding at end of previous year	432	
3.5 Changes in stocks of oil and oil products held overseas by the UK reporting company (see note 3.5)	a. Increase	441	
	b. Decrease	442	

Name of person to be contacted should this be necessary ------------------------
BLOCK CAPITALS PLEASE

Position in company ------------------------ Signature ------------------------

Tel No/Ext ------------------ Telex ------------------ Date ------------------

SC406C

NOTES ON COMPLETING FORM OT/OIL/IN/86

GENERAL

For the purpose of this return:

a the definition of a subsidiary is that given in Section 736 of the Companies Act 1985; (this is the same definition as that previously given in Section 154 of the Companies Act 1948).

b an overseas fellow subsidiary company or branch is a concern which has the same ultimate parent company as yourself but which possesses no direct shareholding in your company. If your immediate parent company is itself a subsidiary of another overseas company the ultimate overseas parent should be treated as a fellow subsidiary company.

c where UK members of the group are immediately controlled through an overseas incorporated but **dual resident holding company**, and all of this company's activities are restricted to the UK, such a holding company should be regarded as resident in the United Kingdom for balance of payments purposes. Its figures should therefore be reported as if it was in the position of a UK holding company.

d. a UK incorporated company is an associate of an overseas company if either the overseas company's interest in the UK company is effectively that of a partner in a joint venture or consortium, or the overseas company's direct interest in the UK company is long-term and substantial (i.e not less than 20% but not more than 50% of the equity voting rights) and, having regard to the disposition of the other shareholdings, the overseas company is in a position to exercise a significant influence over the UK company

e United Kingdom covers Great Britain, Northern Ireland, Isle of Man, the Channel Islands, and that part of the Continental Shelf which is defined by international convention as belonging to the UK(UKCS) The Republic of Ireland is regarded as an overseas country and any transactions with a parent or a related concern incorporated or registered in that country should be included in this return

1. COUNTRY IN WHICH OVERSEAS PARENT IS INCORPORATED OR REGISTERED

If you have more than one overseas investor please enter largest shareholder in first column and smaller investor in second column

2. OVERSEAS DIRECT INVESTMENT IN AND EARNINGS OF UK COMPANY

2.1 The share of profits should be determined in accordance with the overseas company's proportionate holding of the reporting company's equity share capital Refer to us if you have any difficulties in determining this proportion (e.g various classes of voting shares) For balance of payments accounts profits are measured net of **accrued** taxation only, rather than the sum of current and deferred taxation which is more usual company accounting procedure. If you have a subsidiary or associate operating in the UKCS do not report its profits in this question. Such profits should be reported on a separate inquiry (see covering letter)

2.2 Interest paid should be net of any UK withholding tax you deducted. Interest received should be net of any tax paid overseas. If amounts received exceed those paid then enter figures as minus (−) Interest which has accrued but not yet been paid/received should not be reported in this question - it will be reflected in question 2.6

2.3 Include dividends on preference shares together with any dividends which you may have received from subsidiaries or associates operating in the UKCS and which you have passed on to your overseas parent. Exclude dividends declared but not yet remitted.

2.4 If you have permission from the Board of Inland Revenue to pay tax credits direct to the overseas company under the UK double taxation agreements with Denmark, Luxembourg, the Netherlands, Finland, Switzerland, Sweden, Norway, Canada, and the USA please show the amount paid in question 2.4.

2.5 **Include** transactions in quoted and unquoted shares and debentures of the reporting company plus loan stock listed on the UK Stock Exchange; purchases and sales of both existing and new issues of the reporting company's securities; acquisitions made against equipment, manufacturing rights, securities etc. supplied by the overseas company; acquisitions arising from the capitalisation of inter-company current account balances or unlisted loans (with a corresponding reduction in the net amount due from you in question 2.6), purchases where an overseas company's shares are issued in exchange for your equity (value this at the market value of the overseas company's shares on day of purchase); and capital contributions from overseas company. Also include in 2.5a any capital injections which are not intended for repayment and on which no interest is payable.

Exclude bonus issues of shares arising from the capitalisation of reserves, and transactions in unlisted loan stock (the latter should be included against question 2.6).

2.6 **Include** unquoted loan stock.

Exclude any dividends declared but not remitted and any bonus issues of shares arising from the capitalisation of reserves.

Any acquisitions of shares or debentures or listed loan stock arising from the capitalisation of inter-company current account balances or loans should be included in question 2.5a with a corresponding fall shown in the net amount due to you in question 2 6

2 7 If these figures are not available please estimate, or if no estimate is possible enter "N/A" If zero please enter "NIL"

2.8 Please give an estimate of the total Corporation Tax for which liability was incurred **during** the reporting year.

3. OTHER OVERSEAS PAYMENTS AND RECEIPTS

It may not be possible to split payments or receipts between royalties and other transactions (e.g. where amounts include both royalties and other services such as advertising, market research, use of distribution and marketing facilities, etc.) In this case enter the total amount in question 3.1 if royalties account for half or more, or in question 3.2 if royalties account for less than half.

3.1 Include

a. Royalties on printed matter, sound recordings, performing rights, refining process etc., patents, licences, trademarks, designs, copyrights, mining and quarrying mineral rights (but not oil royalties), manufacturing rights, research and development, technical know-how and assistance.

b. Any lump-sum settlements made in year.

c. Amounts payable to overseas agents for charges (less discounts) for services rendered in connection with royalties and similar transactions.

d. Total value of royalties etc. which you collect from, or pay to, overseas residents even though you may be acting as an agent.

Exclude

a. Oil royalties.

b. Royalties, etc., included in the invoiced price of goods which you bought or sold.

c. Royalties, rentals and receipts for films and TV programmes.

d. Payments for UK imports and exports, and their shipping and freight costs; and for films and TV programmes.

3.2 Include

Merchanting of oil and oil products in respect of third-country trade, management and head office expenses, accountants' fees, directors' fees, agents' commission, advertising, consultancy and design fees; legal services, trustee fees; insurance premiums and claims, etc.

Exclude payments for UK imports and exports, and their shipping and freight costs; and for films and TV programmes

3.3 Include trade credit received by the UK reporting group or company in respect of purchases from non-UK residents of goods and services, (including trade credit received from non-UK financial institutions). Subtract any advance and progress payments made.

Exclude trade credit received from overseas members of the group - such credit should be reflected in inter-company balances in question 2.6.

E.g. If a tanker of crude oil is purchased by the UK company from a non-UK resident outside the group in the second week of December under a contract requiring payment in 30 days, then the credit received on the purchase will still be outstanding at the end of the year.

3.4 Include trade credit extended by the UK reporting group or company in respect of sales to non-UK residents of goods and services. Subtract any advance and progress payments received.

Exclude trade credit extended to overseas members of the group - such credit should be reflected in inter-company balances in question 2.6.

E.g If an oil product is sold by the UK company to a non-UK resident outside the group in the second week of December under a contract requiring payment in 30 days, the credit extended on the sale will still be outstanding at the end of the year.

3.5 Exclude stocks of goods held through overseas subsidiary and associate companies.

PLEASE KEEP THESE NOTES FOR REFERENCE

SC406E

A voluntary inquiry conducted by the Government Statistical Service
IN CONFIDENCE

DEPARTMENT OF TRADE AND INDUSTRY
Statistics Division 2
Room 252, Sanctuary Buildings
20 Great Smith Street
LONDON SW1P 3DB

Telephone: 01-215 6235/6293

Date February 1987

	Ref	Sub	YE	UKInd	Tp
1			86	21	

CA/OIL/IN/86

← Please quote this reference number when writing to us

Please correct any errors in name or address

ASSETS AND LIABILITIES AT END 1986
OF OVERSEAS OIL COMPANIES OPERATING ONSHORE IN THE UK

Dear Contributor

We are conducting this annual inquiry to obtain information on the value of overseas oil companies' investment in their subsidiaries and associates which operate onshore in the United Kingdom; (any assets and liabilities relating to the interests of companies in the UK sector of the Continental Shelf should be reported on a separate inquiry). This inquiry is the only source of such data and is used in compiling the UK balance of payments statistics. There is also considerable demand outside government for comprehensive information on the value of overseas investment.

All information you supply will be treated as confidential. It will not be disclosed outside this office and will be used only for compiling totals which will not disclose the figures for individual contributors.

This form should be returned even where there is little or nothing to report against some of the questions. I should like to emphasise that if you do not have precise figures readily available we are happy to accept your best estimates.

I would be grateful if you could return the completed form to us by 22 May 1987, or as soon as possible afterwards.

If you have any difficulties my staff will be pleased to help you. The telephone extensions of those dealing with this inquiry are shown above.

Yours faithfully

P H Richardson

P H RICHARDSON
Chief Statistician

SC402A

PLEASE RETURN THIS FORM BY 22 MAY 1987

For official use

Ref	Sub	Form
9		86

	1	2
601		
605		

IMPORTANT **Please refer to the enclosed notes before completing this form.**
If you do not have precise figures readily available give the best estimates you can.
Please show all values in sterling rounded in £ million to one decimal place(e.g. £53,440,000 should be entered as 53.4). Use the same rates of conversion as in your own accounts.
If you have more than one overseas investor please enter the largest shareholder in the first column and the smaller investor in the second column.

DATE OF RETURN Please give details for 31 December 1986

1. GENERAL PARTICULARS (see note 1)

		1	2
1.1 Name of immediate parent or associate company			
1.2 Country in which immediate parent or associate company is incorporated			
1.3 Country in which ultimate parent company is incorporated (note 1e)			
1.4 Type of business of UK reporting company			
1.5 Percentage of issued share capital of UK reporting company owned by the immediate overseas parent or associate company	602	%	%

UK SUBSIDIARY COMPANY (Associate companies –see note 1c– need complete question 5 only)
Consolidated balance sheet at book value of UK company including all subsidiary companies whether incorporated in the UK or overseas as at 31.12.86.

		£ million	£ million
2. ASSETS			
2.1 Fixed assets (see note 2a)	611		
2.2 Current assets (see note 2b)	613		
2.3 Total assets (2.1 plus 2.2)	606		
3. LIABILITIES			
3.1 Current liabilities (see note 3a)	614		
3.2 Long-term liabilities (see note 3b)			
(a) owing to the immediate and ultimate overseas parent companies and overseas fellow subsidiaries (see note 3c)	615		
(b) other (see note 3d)	616		
3.3 Deferred tax and any long-term provisions or reserves (see note 3e)	617		
3.4 Minority shareholders' interests in any of the UK company's subsidiaries consolidated in the above figures	610		
3.5 Value of share capital and reserves of UK company (including preference shares)			
(a) directly attributable to immediate overseas parent company (see note 3f)	635		
(b) attributable to other shareholders	620		
3.6 Total liabilities (sum of 3.1 to 3.5) (Should equal total assets at 2.3)	607		
4. ADDITIONAL ITEMS			
Inter-company account balance with immediate and ultimate overseas parents and overseas fellow subsidiaries included in:			
4.1 current assets at 2.2 above	618		
4.2 current liabilities at 3.1 above	619		

5. UK ASSOCIATE COMPANY (see note 1c)

Value of reporting company's share capital (including preference shares) and reserves directly owned by immediate overseas company plus loans and net current account balances due to overseas company	627		

Name of person to be contacted should this be necessary ____________________
BLOCK CAPITALS PLEASE

Position in company ____________________ Signature ____________________

Telephone No/Ext ____________________ Telex ____________________ Date ____________________

SC402B

PLEASE KEEP THIS COPY FOR REFERENCE

ASSETS AND LIABILITIES AT END 1986 OF OVERSEAS OIL COMPANIES OPERATING ONSHORE IN THE UK

IMPORTANT **Please refer to the enclosed notes before completing this form. If you do not have precise figures readily available give the best estimates you can. Please show all values in sterling rounded in £ million to one decimal place(e.g. £53,440,000 should be entered as 53.4). Use the same rates of conversion as in your own accounts. If you have more than one overseas investor please enter the largest shareholder in the first column and the smaller investor in the second column.**

DATE OF RETURN **Please give details for 31 December 1986**

1. GENERAL PARTICULARS (see note 1)			
1.1 Name of immediate parent or associate company			
1.2 Country in which immediate parent or associate company is incorporated			
1.3 Country in which ultimate parent company is incorporated (note 1e)			
1.4 Type of business of UK reporting company			
1.5 Percentage of issued share capital of UK reporting company owned by the immediate overseas parent or associate company	602	%	%
UK SUBSIDIARY COMPANY (Associate companies –see note 1c– need complete question 5 only) Consolidated balance sheet at book value of UK company including all subsidiary companies whether incorporated in the UK or overseas as at 31 12 86		£ million	£ million
2. ASSETS			
2.1 Fixed assets (see note 2a)	611		
2.2 Current assets (see note 2b)	613		
2.3 Total assets (2.1 plus 2.2)	606		
3. LIABILITIES			
3.1 Current liabilities (see note 3a)	614		
3.2 Long-term liabilities (see note 3b)			
(a) owing to the immediate and ultimate overseas parent companies and overseas fellow subsidiaries (see note 3c)	615		
(b) other (see note 3d)	616		
3.3 Deferred tax and any long-term provisions or reserves (see note 3e)	617		
3.4 Minority shareholders' interests in any of the UK company's subsidiaries consolidated in the above figures	610		
3.5 Value of share capital and reserves of UK company (including preference shares)			
(a) directly attributable to immediate overseas parent company (see note 3f)	635		
(b) attributable to other shareholders	620		
3.6 Total liabilities (sum of 3.1 to 3.5) (Should equal total assets at 2.3)	607		
4. ADDITIONAL ITEMS Inter-company account balance with immediate and ultimate overseas parents and overseas fellow subsidiaries included in:			
4.1 current assets at 2.2 above	618		
4.2 current liabilities at 3.1 above	619		
5. UK ASSOCIATE COMPANY (see note 1c) Value of reporting company's share capital (including preference shares) and reserves directly owned by immediate overseas company plus loans and net current account balances due to overseas company	627		

Name of person to be contacted should this be necessary ______________________
BLOCK CAPITALS PLEASE

Position in company ______________________ Signature ______________________

Telephone No/Ext ______________________ Telex ______________ Date ______________

SC402D

NOTES ON COMPLETING FORM CA/OIL/IN/86

1. GENERAL PARTICULARS

For the purpose of this return:

(a) a United Kingdom company is normally one incorporated in Great Britain, Northern Ireland, Isle of Man or the Channel Islands. However, where UK members of the group are immediately controlled through an overseas incorporated but dual resident holding company, and all of this company's activities are restricted to the UK, such a holding company should be regarded as resident in the UK for balance of payments purposes. Its figures should therefore be reported as if it were in the position of a UK holding company.

(b) the definition of a subsidiary company is that given in Section 736 of the Companies Act 1985; (this is the same definition as that previously given in section 154 of the Companies Act 1948). Subsidiary companies should not complete question 5.

(c) a UK incorporate company is an associate of an overseas company if either the overseas company's interest in the UK is effectively that of a partner in a joint venture or consortium, or the overseas company's direct interest in the UK company is long-term and substantial (ie not less than 20% but not more than 50% of the equity voting rights) and, having regard to the disposition of the other shareholdings, the overseas company is in a position to exercise a significant influence over the UK company. Associate companies need complete question 5 only.

(d) an overseas fellow subsidiary company is a concern overseas which has the same ultimate parent company as yourself but which possesses no direct shareholding in your company.

(e) if your immediate parent or associate company is a subsidiary of another company please state the country in which the ultimate parent company is incorporated.
If your immediate parent or associate company is not a subsidiary of another company write 'NONE' in the box.

2. ASSETS

(a) Include all tangible and intangible fixed assets (including goodwill) and investments in associates held by the UK reporting company after allowing for depreciation plus investment in securities Deduct any government grants on these assets held as deferred credits or reserves.

(b) Include stocks, debtors, short-term loans, bank balances, cash and long-term loans by the UK company to immediate and ultimate overseas parent companies and to overseas fellow subsidiary companies.

3. LIABILITIES

(a) Include bank overdrafts and short-term loans, trade and other creditors, dividends and interest due, current taxation, short-term borrowing and other provisions of a current character. Exclude deferred taxation.

(b) Include any debentures, long-term loans, or other loans not included in current liabilities at 3.1.

(c) Include any long-term loans, etc, due to your immediate and ultimate parent companies and overseas fellow subsidiary companies including those in different overseas countries from your immediate parent company.

(d) Include any long-term loans, etc, raised in the UK and overseas other than those from your immediate and ultimate parent companies or from overseas fellow subsidiaries in the group.

(e) Exclude any reserves which are part of the shareholders' interest.

(f) Include share capital and reserves attributable to the immediate parent company and any held directly by the ultimate parent company and fellow subsidiaries. However, if you also have an associated investing company with the same ultimate parent as yourself for which you have given separate figures in the second column, you should enter its shareholding in the first column at 3.5(b) and also include it in the second column at question 5.

PLEASE KEEP THESE NOTES AND THE COPY OVERLEAF FOR REFERENCE

SC402C

A voluntary inquiry conducted by the Government Statistical Service
IN CONFIDENCE

DEPARTMENT OF TRADE AND INDUSTRY
Statistics Division 2
Room 252, Sanctuary Buildings
20 Great Smith Street
LONDON SW1P 3DB

Telephone: 01-215 6235/6293

Date February 1987

OT/OIL/NS/86

← Please quote this reference number when writing to us

ANNUAL INQUIRY INTO OVERSEAS TRANSACTIONS OF OIL COMPANIES HAVING INTERESTS IN THE UKCS IN 1986

Dear Contributor

The purpose of this inquiry is to obtain up-to-date statistical information about various transactions between foreign owned oil companies involved in activities in the UKCS and their parent company overseas and overseas fellow subsidiaries and branches for use in compiling our country's balance of payments statistics. Similar transactions relating to UK onshore activities should be reported on a separate inquiry. The information is needed by government in deciding its economic policy as well as its commercial policy in relation to other countries. This inquiry is the only source of such information and without it the accuracy of the UK balance of payments statistics will be greatly reduced.

The current form is enclosed and I would be most grateful if you would complete it and return it to me preferably not later than 22 May this year.

All information supplied will be treated as confidential. Only those persons directly employed on collecting, processing and interpreting the information will have access to the particulars of individual returns, which will not be disclosed outside this small group of officials.

If you have any difficulties or would like any further information about this particular inquiry, my staff will be pleased to help you. The telephone extensions of those dealing with this inquiry are shown above.

Yours faithfully

P H Richardson.

P H RICHARDSON
Chief Statistician

SC407A

NOTES ON COMPLETING FORM OT/OIL/NS/86

GENERAL

For the purpose of this return:

a. the definition of a subsidiary is that given in Section 736 of the Companies Act 1985; (this is the same definition as that previously given in Section 154 of the Companies Act 1948).

b. an overseas "fellow" subsidiary company or branch is a concern which has the same ultimate parent company as yourself bu which possesses no direct shareholding in your company. If your immediate parent company is itself a subsidiary of another overseas company the ultimate overseas parent should be treated as a fellow subsidiary company.

c where UK members of the group are immediately controlled through an overseas incorporated but **dual resident holding company**, and all of this company's activities are restricted to the UK, such a holding company should be regarded as resident in the United Kingdom for balance of payments purposes. Its figures should therefore be reported as if it was in the position of a UK holding company.

d. a UK incorporated company is an associate of an overseas company if either the overseas company's interest in the UK company is effectively that of a partner in a joint venture or consortium, or the overseas company's direct interest in the UK company is long-term and substantial (i.e not less than 20% but not more than 50% of the equity voting rights) and, having regard to the disposition of the other shareholdings, the overseas company is in a position to exercise a significant influence over the UK company.

e. United Kingdom covers Great Britain, Northern Ireland, Isle of Man, the Channel Islands, and that part of the Continental Shelf which is defined by international convention as belonging to the UK(UKCS). The Republic of Ireland is regarded as an overseas country and any transactions with a parent or a related concern incorporated or registered in that country should be included in this return.

1. COUNTRY IN WHICH OVERSEAS PARENT IS INCORPORATED OR REGISTERED

If you have more than one overseas investor please enter largest shareholder in first column and smaller investor in second column.

2. OVERSEAS DIRECT INVESTMENT IN AND EARNINGS OF UK COMPANY

2.1 The share of profits should be determined in accordance with the overseas company's proportionate holding of the reporting company's equity share capital. Refer to us if you have any difficulties in determining this proportion (e.g. various classes of voting shares). For balance of payments accounts profits are measured net of **accrued** taxation only, rather than the sum of current and deferred taxation which is more usual company accounting procedure If you have a subsidiary or associate operating onshore in the UK do not report its profits in this question. Such profits should be reported on a separate inquiry (see covering letter.)

2.2 Interest paid should be net of any UK withholding tax you deducted. Interest received should be net of any tax paid overseas. If amounts received exceed those paid then enter figures as minus (–). Interest which has accrued but not yet been paid/received should not be reported in this question – it will be reflected in question 2.6.

2.3 Include dividends on preference shares together with any dividends which you may have received from subsidiaries or associates operating onshore in the UK, and which you have passed on to your overseas parent. Exclude dividends declared but not yet remitted.

2.4 If you have permission from the Board of Inland Revenue to pay tax credits direct to the overseas company under the UK double taxation agreements with Denmark, Luxembourg, the Netherlands, Finland, Switzerland, Sweden, Norway, Canada and the USA please show the amount paid in question 2.4.

2.5 **Include** transactions in quoted and unquoted shares and debentures of the reporting company plus loan stock listed on the UK Stock Exchange; purchases and sales of both existing and new issues of the reporting company's securities; acquisitions made against equipment, manufacturing rights, securities etc. supplied by the overseas company; acquisitions arising from the capitalisation of inter-company current account balances or unlisted loans (with a corresponding reduction in the net amount due from you in question 2.6); purchases where an overseas company's shares are issued in exchange for your equity (value this at the market value of the overseas company's shares on day of purchase); and capital contributions from overseas company. Also include in question 2.5a any capital injections which are not intended for repayment and on which no interest is payable.

Exclude bonus issues of shares arising from the capitalisation of reserves, and transactions in unlisted loan stock (the latter should be included against question 2.6).

2.6 **Include** unquoted loan stock.

Exclude any dividends declared but not remitted and any bonus issues of shares arising from the capitalisation of reserves.

Any acquisitions of shares or debentures or listed loan stock arising from the capitalisation of inter-company current account balances or loans should be included in question 2.5a with a corresponding fall shown in the net amount due to you in question 2.6.

2.7 If these figures are not available please estimate, or if no estimate is possible enter "N/A". If zero, please enter "NIL"

2.8 Please give estimates of APRT, PRT and CT for which liability was incurred during the reporting year. If it is not possible to split Petroleum Revenue Tax between APRT and the remaining PRT liability then please record the total liability against part a. of the question and write in part b "included above".

3. OTHER OVERSEAS TRANSACTIONS

3.1 **Include** trade credit extended by the UK reporting group or company in respect of sales of oil to non-UK residents. Subtract any advance and progress payments received.

Exclude trade credit extended to overseas members of the group – such credit should be reflected in inter-company balances in question 2.6.

E.g If a tanker of oil is sold by the UK company to a non-UK resident outside the group in the second week of December under a contract requiring payment in thirty days, the credit extended on the sale will still be outstanding at the end of the year.

3.2 Exclude stocks of goods held through overseas subsidiary and associate companies.

PLEASE KEEP THESE NOTES FOR REFERENCE SC407B

A voluntary inquiry conducted by the Government Statistical Service

IN CONFIDENCE

DEPARTMENT OF TRADE AND INDUSTRY
Statistics Division 2
Room 252, Sanctuary Buildings
20 Great Smith Street
LONDON SW1P 3DB

Telephone: 01-215 6235/6293

Date February 1987

Ref	Sub	YE	UKInd	Tp
1		86	22	

OT/OIL/NS/86

← Please quote this reference number when writing to us

For official use

Ref	Sub	Form
10	86	

	1	2
161		

Please correct any errors in name or address

ANNUAL INQUIRY INTO OVERSEAS TRANSACTIONS OF OIL COMPANIES HAVING INTERESTS IN THE UKCS IN 1986

IMPORTANT Please refer to the enclosed notes before completing this form.
If you do not have precise figures readily available give the best estimates you can.
Please show all values in sterling rounded in £ million to one decimal place (e.g. £53,440,000 should be entered as 53.4). Use the same rates of conversion as in your own accounts.

DATE OF RETURN : Please give details for 31 December 1986

1. COUNTRY IN WHICH OVERSEAS PARENT OR ASSOCIATE SHAREHOLDER IS INCORPORATED OR REGISTERED (see note 1) .... [|]

2. OVERSEAS DIRECT INVESTMENT IN AND EARNINGS OF UK COMPANY

Item		Box	£ million	£ million
2.1 Overseas company's share of the UK reporting company's consolidated profit or loss (whether remitted or not) after allowing for provision for depreciation, accrued tax and extraordinary items (see note 2.1)	a. Profit	162		
	b. Loss	163		
2.2 Interest paid to, less interest received from, overseas parent company and overseas fellow subsidiaries and branches (see note 2.2)		164		
2.3 Dividends paid to overseas parent company (see note 2.3)		165		
2.4 UK tax credits paid directly by you to overseas parent company in respect of dividends (see note 2.4)		166		
2.5 Purchases and sales by the overseas parent company of the reporting company's shares (including preference shares) debentures and listed loan stock (see note 2.5)	a. Purchases	173		
	b Sales	174		
2.6 Net total of inter-company current account balances, long-term loans (other than debentures and listed loan stock) and short-term loans outstanding between UK reporting company and the overseas company, or overseas fellow subsidiaries, branches and associates (see note 2.6)	a. Net amount due from UK company at end of current year	176		
	OR b. Net amount due to UK company at end of current year	177		
	AND c Net amount due from UK company at end of previous year	178		
	OR d. Net amount due to UK company at end of previous year	179		

SC407C

PLEASE RETURN THIS FORM BY 22 MAY 1987

			£ million	£ million
2.7 Net gain or loss resulting from exchange rate movements made by the reporting company on the balances reported in question 2.6 (see note 2.7)	a. Gain	168		
	OR b. Loss	169		

2.8 Taxation accrued during year (see note 2.8)	a. Advance petroleum revenue tax	412		
	b. Petroleum revenue tax	411		
	c. Corporation tax	414		

3. OTHER OVERSEAS TRANSACTIONS

3.1 Total net trade credit extended by UK reporting group or company in respect of oil sales to non-UK residents outside the group (including credit extended on oil bought and sold outside the UK) (see note 3.1)	a. Net credit extended outstanding at end of current year	431		
	b. Net credit extended outstanding at end of previous year	432		
3.2 Changes in stocks of oil and oil products held overseas by the UK reporting company (see note 3.2)	a. Increase	441		
	b. Decrease	442		

Name of person to be contacted should this be necessary ---------------------------------------
BLOCK CAPITALS PLEASE

Position in company ------------------------------ Signature ------------------------------

Tel No/Ext -------------------- Telex -------------------- Date --------------------

SC407D

A voluntary inquiry conducted by the Government Statistical Service

IN CONFIDENCE

DEPARTMENT OF TRADE AND INDUSTRY
Statistics Division 2
Room 252, Sanctuary Buildings
20 Great Smith Street
LONDON SW1P 3DB

Telephone: 01-215 6235/6293

Date February 1987

Ref		Sub	YE	UKInd	Tp
1			86	22	

CA/OIL/NS/86

← Please quote this reference number when writing to us

Please correct any errors in name or address

ASSETS AND LIABILITIES AT END 1986
OF OVERSEAS OIL COMPANIES WITH INTERESTS IN THE UK SECTOR OF THE CONTINENTAL SHELF

Dear Contributor

We are conducting this annual inquiry to obtain information on the value of overseas oil companies' investment in their subsidiaries and associates with UKCS interests; (any assets and liabilities relating to companies' onshore activities should be reported on a separate inquiry). This inquiry is the only source of such data and is used in compiling the UK balance of payments statistics. There is also considerable demand outside government for comprehensive information on the value of overseas investment.

All information you supply will be treated as confidential. It will not be disclosed outside this office and will be used only for compiling totals which will not disclose the figures for individual contributors.

This form should be returned even where there is little or nothing to report against some of the questions. I should like to emphasise that if you do not have precise figures readily available we are happy to accept your best estimates.

I would be grateful if you could return the completed form to us by 22 May 1987, or as soon as possible afterwards.

If you have any difficulties my staff will be pleased to help you. The telephone extensions of those dealing with this inquiry are shown above.

Yours faithfully

P H Richardson

P H RICHARDSON
Chief Statistician

SC403A

PLEASE RETURN THIS FORM BY 22 MAY 1987

For official use

	Ref	Sub	Form
9		86	

	1	2
601		
605		

IMPORTANT **Please refer to the enclosed notes before completing this form. If you do not have precise figures readily available give the best estimates you can. Please show all values in sterling rounded in £ million to one decimal place(e.g. £53,440,000 should be entered as 53.4). Use the same rates of conversion as in your own accounts. If you have more than one overseas investor please enter the largest shareholder in the first column and the smaller investor in the second column.**

DATE OF RETURN **Please give details for 31 December 1986**

1. GENERAL PARTICULARS (see note 1)

1.1 Name of immediate parent or associate company			
1.2 Country in which immediate parent or associate company is incorporated			
1.3 Country in which ultimate parent company is incorporated (note 1e)			
1.4 Type of business of UK reporting company			
1.5 Percentage of issued share capital of UK reporting company owned by the immediate overseas parent or associate company	602	%	%

UKCS SUBSIDIARY COMPANY (Associate companies –see note 1c– need complete question 5 only)
Consolidated balance sheet at book value of UK company including all subsidiary companies whether incorporated in the UK or overseas as at 31.12.86.

		£ million	£ million
2. ASSETS			
2.1 Fixed assets (see note 2a)	611		
2.2 Current assets (see note 2b)	613		
2.3 Total assets (2.1 plus 2.2)	606		
3. LIABILITIES			
3.1 Current liabilities (see note 3a)	614		
3.2 Long-term liabilities (see note 3b)			
(a) owing to the immediate and ultimate overseas parent companies and overseas fellow subsidiaries (see note 3c)	615		
(b) other (see note 3d)	616		
3.3 Deferred tax and any long-term provisions or reserves (see note 3e)	617		
3.4 Minority shareholders' interests in any of the UK company's subsidiaries consolidated in the above figures	610		
3.5 Value of share capital and reserves of UK company (including preference shares)			
(a) directly attributable to immediate overseas parent company (see note 3f)	635		
(b) attributable to other shareholders	620		
3.6 Total liabilities (sum of 3.1 to 3.5) (Should equal total assets at 2.3)	607		
4. ADDITIONAL ITEMS			
Inter-company account balance with immediate and ultimate overseas parents and overseas fellow subsidiaries included in:			
4.1 current assets at 2.2 above	618		
4.2 current liabilities at 3.1 above	619		
5. UKCS ASSOCIATE COMPANY (see note 1c)			
Value of reporting company's share capital (including preference shares) and reserves directly owned by immediate overseas company plus loans and net current account balances due to overseas company	627		

Name of person to be contacted should this be necessary ______________________
BLOCK CAPITALS PLEASE

Position in company ______________________ Signature ______________________

Telephone No/Ext ______________________ Telex ______________ Date ______________

SC403B

PLEASE KEEP THIS COPY FOR REFERENCE

ASSETS AND LIABILITIES AT END 1986 OF OVERSEAS OIL COMPANIES WITH INTERESTS IN THE UK SECTOR OF THE CONTINENTAL SHELF

IMPORTANT **Please refer to the enclosed notes before completing this form. If you do not have precise figures readily available give the best estimates you can. Please show all values in sterling rounded in £ million to one decimal place(e.g. £53,440,000 should be entered as 53.4).** Use the same rates of conversion as in your own accounts. If you have more than one overseas investor please enter the largest shareholder in the first column and the smaller investor in the second column.

DATE OF RETURN Please give details for 31 December 1986

1. GENERAL PARTICULARS (see note 1)			
1.1 Name of immediate parent or associate company			
1.2 Country in which immediate parent or associate company is incorporated			
1.3 Country in which ultimate parent company is incorporated (note 1e)			
1.4 Type of business of UK reporting company			
1.5 Percentage of issued share capital of UK reporting company owned by the immediate overseas parent or associate company	602	%	%

UKCS SUBSIDIARY COMPANY (Associate companies -see note 1c- need complete question 5 only)
Consolidated balance sheet at book value of UK company including all subsidiary companies whether incorporated in the UK or overseas as at 31.12.86.

		£ million	£ million
2. ASSETS			
2.1 Fixed assets (see note 2a)	611		
2.2 Current assets (see note 2b)	613		
2.3 Total assets (2.1 plus 2.2)	606		
3. LIABILITIES			
3.1 Current liabilities (see note 3a)	614		
3.2 Long-term liabilities (see note 3b)			
(a) owing to the immediate and ultimate overseas parent companies and overseas fellow subsidiaries (see note 3c)	615		
(b) other (see note 3d)	616		
3.3 Deferred tax and any long-term provisions or reserves (see note 3e)	617		
3.4 Minority shareholders' interests in any of the UK company's subsidiaries consolidated in the above figures	610		
3.5 Value of share capital and reserves of UK company (including preference shares)			
(a) directly attributable to immediate overseas parent company (see note 3f)	635		
(b) attributable to other shareholders	620		
3.6 Total liabilities (sum of 3.1 to 3.5) (Should equal total assets at 2.3)	607		
4. ADDITIONAL ITEMS			
Inter-company account balance with immediate and ultimate overseas parents and overseas fellow subsidiaries included in:			
4.1 current assets at 2.2 above	618		
4.2 current liabilities at 3.1 above	619		
5. UKCS ASSOCIATE COMPANY (see note 1c)			
Value of reporting company's share capital (including preference shares) and reserves directly owned by immediate overseas company plus loans and net current account balances due to overseas company	627		

Name of person to be contacted should this be necessary ______________________
BLOCK CAPITALS PLEASE

Position in company ______________________ Signature ______________________

Telephone No/Ext ______________________ Telex ______________ Date ______________

SC403D

NOTES ON COMPLETING FORM CA/OIL/NS/86

1. GENERAL PARTICULARS

For the purpose of this return:

(a) a United Kingdom company is normally one incorporated in Great Britain, Northern Ireland, Isle of Man or the Channel Islands. However, where UK members of the group are immediately controlled through an overseas incorporated but dual resident holding company, and all of this company's activities are restricted to the UK, such a holding company should be regarded as resident in the UK for balance of payments purposes. Its figures should therefore be reported as if it were in the position of a UK holding company.

(b) the definition of a subsidiary company is that given in Section 736 of the Companies Act 1985; (this is the same definition as that previously given in section 154 of the Companies Act 1948). Subsidiary companies should not complete question 5.

(c) a UK incorporate company is an associate of an overseas company if either the overseas company's interest in the UK is effectively that of a partner in a joint venture or consortium, or the overseas company's direct interest in the UK company is long-term and substantial (ie not less than 20% but not more than 50% of the equity voting rights) and, having regard to the disposition of the other shareholdings, the overseas company is in a position to exercise a significant influence over the UK company. Associate companies need complete question 5 only.

(d) an overseas fellow subsidiary company is a concern overseas which has the same ultimate parent company as yourself but which possesses no direct shareholding in your company.

(e) if your immediate parent or associate company is a subsidiary of another company please state the country in which the ultimate parent company is incorporated.
If your immediate parent or associate company is not a subsidiary of another company write 'NONE' in the box.

2. ASSETS

(a) Include all tangible and intangible fixed assets (including goodwill) and investments in associates held by the UK reporting company after allowing for depreciation plus investment in securities. Deduct any government grants on these assets held as deferred credits or reserves.

(b) Include stocks, debtors, short-term loans, bank balances, cash and long-term loans by the UK company to immediate and ultimate overseas parent companies and to overseas fellow subsidiary companies.

3. LIABILITIES

(a) Include bank overdrafts and short-term loans, trade and other creditors, dividends and interest due, current taxation, short-term borrowing and other provisions of a current character. Exclude deferred taxation.

(b) Include any debentures, long-term loans, or other loans not included in current liabilities at 3.1.

(c) Include any long-term loans, etc, due to your immediate and ultimate parent companies and overseas fellow subsidiary companies including those in different overseas countries from your immediate parent company.

(d) Include any long-term loans, etc, raised in the UK and overseas other than those from your immediate and ultimate parent companies or from overseas fellow subsidiaries in the group.

(e) Exclude any reserves which are part of the shareholders' interest.

(f) Include share capital and reserves attributable to the immediate parent company and any held directly by the ultimate parent company and fellow subsidiaries. However, if you also have an associated investing company with the same ultimate parent as yourself for which you have given separate figures in the second column, you should enter its shareholding in the first column at 3.5(b) and also include it in the second column at question 5.

PLEASE KEEP THESE NOTES AND THE COPY OVERLEAF FOR REFERENCE

SC403C

A voluntary inquiry conducted by the Government Statistical Service
IN CONFIDENCE

DEPARTMENT OF TRADE AND INDUSTRY
Statistics Division 2
Room 252, Sanctuary Buildings
20 Great Smith Street
LONDON SW1P 3DB

Telephone 01-215 6235/6293

Date February 1987

	Ref	Sub	YE	UKInd	Tp
1			86	22	

OT/OIL/NS/86B

← Please quote this reference number when writing to us

Please correct any errors in name or address

ANNUAL INQUIRY INTO TRANSACTIONS BETWEEN UNITED KINGDOM BRANCHES OF OIL COMPANIES AND THEIR OVERSEAS HEAD OFFICE IN 1986

Dear Contributor

The purpose of this inquiry is to obtain up-to-date statistical information about various transactions between UK branches of oil companies and their overseas head office and overseas fellow subsidiaries and branches for use in compiling our country's balance of payments statistics. The information is needed by government in deciding its economic policy as well as its commercial policy in relation to other countries. This inquiry is the only source of such information and without it the accuracy of the UK balance of payments statistics will be greatly reduced.

The current form is overleaf and I would be most grateful for your co-operation in completing and returning it to me preferably not later than 22 May this year.

All information supplied will be treated as confidential. Only those persons directly employed on collecting, processing and interpreting the information will have access to the particulars of individual returns, which will not be disclosed outside this small group of officials. The aggregated results will be compiled so as not to reveal the figures of individual contributors.

If you have any difficulties or would like any further information about this particular inquiry, my staff will be pleased to help you. The telephone extensions of those dealing with this inquiry are shown above.

Yours faithfully

P H Richardson.

P H RICHARDSON
Chief Statistician

SC405A

PLEASE RETURN THIS FORM BY 22 MAY 1987

For official use

8	Ref	Sub	86	Form

121 [ ]

IMPORTANT Please refer to the enclosed notes before completing this form.
If you do not have precise figures readily available give the best estimates you can.
Please show all values in sterling rounded in £ million to one decimal place(e.g. £53,440,00 should be entered as 53.4). Use the same rates of conversion as in your own accounts.

DATE OF RETURN: Please give details for 31 December 1986

1. OVERSEAS COUNTRY IN WHICH HEAD OFFICE IS INCORPORATED OR REGISTERED (see note 1)

2. OVERSEAS DIRECT INVESTMENT AND EARNINGS.

£ million

Item	Description	Box	
2.1 Total profit or loss of the branch (whether remitted abroad or not) after allowing for provision for depreciation, United Kingdom accrued tax, and extraordinary items. (see note 2.1).	a. Profit (including any interest payments to overseas head office and to overseas fellow subsidiaries and branches).	122	
	OR		
	b. Loss	123	
2.2 Total value of net worth of branch attributable to overseas head office plus loans and current account balances of branch due to or from overseas fellow subsidiaries and branches. (see note 2.2).	a. Net amount due from UK branch at end of current financial year	125	
	OR		
	b. Net amount due to UK branch at end of current financial year	126	
	AND		
	c. Net amount due from UK branch at end of previous financial year	127	
	OR		
	d. Net amount due to UK branch at end of previous financial year	128	
2.3 Net gain or loss resulting from exchange rate movements made by the branch and reported in question 2.2 above (see note 2.3).	a. Gain	171	
	b. Loss	172	
2.4 Taxation accrued during year (see note 2.4)	a. Advance Petroleum Revenue Tax	416	
	b. Petroleum Revenue Tax	415	
	c. Corporation Tax	418	

3. OTHER OVERSEAS TRANSACTIONS

Item	Description	Box	
Total net trade credit extended by branch in respect of oil sales to non-UK residents outside the group (including credit extended on oil bought and sold outside the UK) (see note 3).	a. Net credit extended outstanding at end of current year	435	
	b. Net credit extended outstanding at end of previous year	436	

Name of person to be contacted should this be necessary ____________________
BLOCK CAPITALS PLEASE

Position in company ____________________ Signature ____________________

Telephone No/Ext ____________________ Telex ____________________ Date ____________________

SC405B

PLEASE KEEP THIS COPY FOR REFERENCE

ANNUAL INQUIRY INTO TRANSACTIONS BETWEEN UNITED KINGDOM BRANCHES OF OIL COMPANIES AND THEIR OVERSEAS HEAD OFFICE IN 1986

IMPORTANT **Please refer to the enclosed notes before completing this form. If you do not have precise figures readily available give the best estimates you can. Please show all values in sterling rounded in £ million to one decimal place(e.g. £53,440,000 should be entered as 53.4). Use the same rates of conversion as in your own accounts.**

DATE OF RETURN: Please give details for 31 December 1986

1. OVERSEAS COUNTRY IN WHICH HEAD OFFICE IS INCORPORATED OR REGISTERED (see note 1) ....

2. OVERSEAS DIRECT INVESTMENT AND EARNINGS.

		Code	£ million
2.1 Total profit or loss of the branch (whether remitted abroad or not) after allowing for provision for depreciation, United Kingdom accrued tax, and extraordinary items. (see note 2.1).	a Profit (including any interest payments to overseas head office and to overseas fellow subsidiaries and branches).	122	
	OR		
	b. Loss	123	
2.2 Total value of net worth of branch attributable to overseas head office plus loans and current account balances of branch due to or from overseas fellow subsidiaries and branches. (see note 2.2).	a. Net amount due from UK branch at end of current financial year	125	
	OR		
	b. Net amount due to UK branch at end of current financial year	126	
	AND		
	c. Net amount due from UK branch at end of previous financial year	127	
	OR		
	d. Net amount due to UK branch at end of previous financial year	128	
2.3 Net gain or loss resulting from exchange rate movements made by the branch and reported in question 2.2 above (see note 2.3).	a. Gain	171	
	b. Loss	172	
2.4 Taxation accrued during year (see note 2.4)	a. Advance Petroleum Revenue Tax	416	
	b. Petroleum Revenue Tax	415	
	c. Corporation Tax	418	

3. OTHER OVERSEAS TRANSACTIONS

		Code	£ million
Total net trade credit extended by branch in respect of oil sales to non-UK residents outside the group (including credit extended on oil bought and sold outside the UK) (see note 3).	a. Net credit extended outstanding at end of current year	435	
	b. Net credit extended outstanding at end of previous year	436	

Name of person to be contacted should this be necessary -
BLOCK CAPITALS PLEASE

Position in company - Signature - - - - - - - - - - - - - - - -

Telephone No/Ext - - - - - - - - - - - - - - - - - - Telex - - - - - - - - - - - - Date - - - - - - - - -

SC405D

NOTES ON COMPLETING FORM OT/OIL/NS/86B

GENERAL

For the purpose of this return:

(a) a "branch" means a "permanent establishment" as defined for the purpose of UK tax and UK double taxation agreements, being a fixed place of business in the UK through which the business of the overseas enterprise is wholly or partly carried on, and which is not a UK company (other questionnaires, OT/OIL/IN and OT/OIL/NS, cover UK companies in which overseas companies hold a significant shareholding);

(b) an overseas "fellow subsidiary company or branch" is a concern overseas which has the same ultimate parent company as yourself. If your overseas head office is itself a subsidiary of another overseas company the ultimate overseas parent should be treated as a fellow subsidiary company.

1. OVERSEAS COUNTRY IN WHICH HEAD OFFICE IS INCORPORATED OR REGISTERED

United Kingdom covers Great Britain, Northern Ireland, the Isle of Man, and the Channel Islands. The Republic of Ireland is regarded as an overseas country and any transactions with a parent or related concern incorporated or registered in that country should be included in this return.

2. OVERSEAS DIRECT INVESTMENT AND EARNINGS

2.1 For balance of payments purposes profits are measured net of **accrued** taxation only, rather than the sum of current and deferred taxation which is more usual company accounting procedure.

2.2 The net worth of the UK branch consists of its fixed assets (net of depreciation) and its investments, plus its net current assets with concerns other than the overseas head office or overseas fellow subsidiaries and branches, less long term loans other than those from the overseas head office or overseas fellow subsidiaries and branches, less accrued tax and deferred credits of the branch, and less any provisions held by the branch. This should equal the unremitted profits and capital reserves of the branch plus loans and net current account indebtedness due to the overseas head office and overseas fellow subsidiaries and branches.

2.3 If these figures are not available please estimate, or if no estimate is possible enter "N/A". If zero, please enter "NIL".

2.4 Please give estimates of APRT, PRT and CT for which liability was incurred during the reporting year. If it is not possible to split Petroleum Revenue Tax between APRT and the remaining PRT liability then please record the total liability against part a of the question and leave part b blank.

3. OTHER OVERSEAS TRANSACTIONS

Include trade credit extended by the UK reporting group or company in respect of sales of oil to non-UK residents. Subtract any advance and progress payments received.

Exclude trade credit extended to overseas members of the group - such credit should be reflected in question 2.2.

E.g. If a tanker of oil is sold by the UK branch to a non-UK resident outside the group in the second week of December under a contract requiring payment in thirty days, the credit extended on the sale will still be outstanding at the end of the year.

PLEASE KEEP THESE NOTES AND THE COPY OVERLEAF FOR REFERENCE

SC405C

A **compulsory** inquiry conducted by the Government Statistical Service
IN CONFIDENCE

DEPARTMENT OF TRADE AND INDUSTRY
Statistics Division 2
Room 252, Sanctuary Buildings
20 Great Smith Street
LONDON SW1P 3DB

Telephone 01-215 6235/6293

Date: February 1987

	Ref	Sub	YE	UKInd	Tp
1			86	22	

CA/OIL/NS/86B

← Please quote this reference number when writing to us

Please correct any errors in name or address

NOTICE UNDER SECTION 1 OF THE STATISTICS OF TRADE ACT 1947

The Secretary of State for Trade and Industry hereby requires you by law to provide the information called for in this inquiry form with the exception of questions 3.3 and 3.4. The information in this form is required for the purposes of the appreciation of economic trends and the provision of a statistical service for government and industry.

Please complete this form and return it to the Department of Trade and Industry by 22 May 1987

N.B. The information given by you under the Act will be treated as confidential in strict accordance with the Act and subject to the further restriction that information about individual businesses will be used and disclosed under Ministerial direction to other government departments for statistical purposes only, except that the names and addresses of individual businesses and their industrial classification may, if a Minister so directs, be made available to Ministers and officials of government departments who need them for carrying out their functions.

ASSETS AND LIABILITIES AT END 1986 OF UK BRANCHES OF OVERSEAS OIL COMPANIES

Dear Contributor

We are conducting this annual inquiry to obtain information on the value of overseas companies' investment in their branches in the United Kingdom. It is the only source of such data and is used in compiling the UK balance of payments statistics.

Please note that questions 3.3 and 3.4 are voluntary and you are not required by law to answer them. However it would be very helpful if you could give this information as it will complete the balance sheet for the branch. Any information given voluntarily will be treated as confidential. It will not be disclosed outside this office and will be used only for compiling totals which will not disclose the figures for individual contributors.

If the total value of overseas investment in your branch (as defined overleaf) is between minus £500,000 and plus £500,000 just tick the box in question 1.1 and return the form to us. You do not have to answer any other questions.

This form should be returned even where there is little or nothing to report against some of the questions. I should like to emphasise that if you do not have precise figures readily available we are happy to accept your best estimates.

I would be grateful if you could return the completed form to us as soon as possible. In any case it should be returned by 22 May 1987.

If you have any difficulties my staff will be pleased to help you. The telephone extensions of those dealing with this inquiry are shown above.

Yours sincerely

P H RICHARDSON
Chief Statistician

SC401A

PLEASE RETURN THIS FORM BY 22 MAY 1987

For official use

	Ref	Sub	Form	
7			86	

601

IMPORTANT Please refer to the enclosed notes before completing this form.
If you do not have precise figures readily available give the best estimates you can.
Please show all values in sterling rounded in £ million to one decimal place (e.g. £53,440, should be entered as 53.4). Use the same rates of conversion as in your own accounts.

DATE OF RETURN: Please give details for 31 December 1986

1. GENERAL PARTICULARS (see note 1)

1.1 If your net worth (see note 3d) is between minus £500,000 and plus £500,000, tick this box then sign and return the form. You do not have to answer any other questions. ☐

1.2 Country in which overseas company is incorporated ....

1.3 Trading name of overseas company if different from trading name in UK.....

1.4 Type of business of UK branch ....

1.5 If you are resident in the UK for tax purposes and have UK subsidiary companies, please list them here.

....

....

AND EITHER include the value of your investment in the UK subsidiary companies in question 2.2 and tick this box .... ☐ 631 2

OR if more convenient, give combined figures for yourself and your UK subsidiaries in each part of questions 2 & 3 and tick this box .... ☐ 631 1

If you do not own all the share capital of a subsidiary company include the minority shareholders' interests in "other liabilities" in question 3.3.

2. ASSETS

		£ million
2.1 Fixed assets after provision for depreciation (see note 2a)	621	
2.2 Investments in company and government securities and loans(see note 2b)	622	
2.3 Current assets (see notes 2b & 2c)	623	

3. LIABILITIES

3.1 Current liabilities (see notes 3a & 3b)	624	
3.2 Long-term liabilities (see notes 3a & 3c)	625	
3.3 Provisions, deferred tax reserves and other liabilities (voluntary question)	626	
3.4 Net worth of UK branch (see note 3d) (voluntary question)	636	

Name of person to be contacted should this be necessary ________________
BLOCK CAPITALS PLEASE

Position in company ________________ Signature ________________

Telephone No/Ext ________________ Telex ________________ Date ________________

SC401B

PLEASE KEEP THIS COPY FOR REFERENCE

ASSETS AND LIABILITIES AT END 1986
OF UK BRANCHES OF OVERSEAS OIL COMPANIES

IMPORTANT Please refer to the enclosed notes before completing this form.
If you do not have precise figures readily available give the best estimates you can.
Please show all values in sterling rounded in £ million to one decimal place (e.g. £53,440,0(should be entered as 53.4). Use the same rates of conversion as in your own accounts.

DATE OF RETURN: Please give details for 31 December 1986

1. GENERAL PARTICULARS (see note 1)

1.1 If your net worth (see note 3d) is between minus £500,000 and plus £500,000, tick this box ☐ then sign and return the form. You do not have to answer any other questions.

1.2 Country in which overseas company is incorporated ....

1.3 Trading name of overseas company if different from trading name in UK ....

1.4 Type of business of UK branch ....

1.5 If you are resident in the UK for tax purposes and have UK subsidiary companies, please list them here.

....

....

AND EITHER include the value of your investment in the UK subsidiary companies in question 2.2 and tick this box .... ☐ 631 2

OR if more convenient, give combined figures for yourself and your UK subsidiaries in each part of questions 2 & 3 and tick this box .... ☐ 631 1

If you do not own all the share capital of a subsidiary company include the minority shareholders' interests in "other liabilities" in question 3.3.

		£ million
2. ASSETS		
2.1 Fixed assets after provicion for depreciation (see note 2a)	621	
2.2 Investments in company and government securities and loans(see note 2b)	622	
2.3 Current assets (see notes 2b & 2c)	623	
3. LIABILITIES		
3.1 Current liabilities (see notes 3a & 3b)	624	
3.2 Long-term liabilities (see notes 3a & 3c)	625	
3.3 Provisions, deferred tax reserves and other liabilities (voluntary question)	626	
3.4 Net worth of UK branch (see note 3d) (voluntary question)	636	

Name of person to be contacted should this be necessary ____________________
BLOCK CAPITALS PLEASE

Position in company ____________________ Signature ____________________

Telephone No/Ext ____________________ Telex ______________ Date ____________

SC401D

NOTES ON COMPLETING FORM CA/OIL/NS/86B

1. GENERAL PARTICULARS

(a) If you are a company incorporated in Great Britain, Northern Ireland, Isle of Man or the Channel Islands please state this in question 1.2 and return the form to us uncompleted.

For the purpose of this return:

(b) A UK branch means a permanent establishment as defined for the purposes of UK tax and UK double taxation agreements, being a place of business in the UK through which the business of the overseas enterprise is wholly or partly carried on and which is not a company incorporated in Great Britain, Northern Ireland, Isle of Man or the Channel Islands.

(c) Overseas members of your group of companies means companies which are:

i. directly or indirectly owned by your company, plus, where the company of which you are a branch is itself a subsidiary, your ultimate parent company and all its directly and indirectly owned subsidiaries; and

ii. incorporated outside Great Britain, Northern Ireland, Isle of Man and the Channel Islands.

(d) For overseas incorporated companies with their head office in the UK, "overseas head office" means the registered office of the company in its country of incorporation.

2. ASSETS

(a) Include all tangible and intangible fixed assets in the UK owned by the company which are used or employed for the business of the branch. Deduct the value of any government grants on fixed assets held as deferred credits.

(b) Exclude from questions 2.2 & 2.3 any loans to and other amounts due from overseas members of your group of companies, and overseas head office and inter-branch balances.

(c) Include stocks, debtors, short-term loans, bank balances, prepayments (deferred assets) and cash.

3. LIABILITIES

(a) Exclude from questions 3.1 and 3.2, loans from and other amounts due to overseas members of your group of companies, overseas head office and inter-branch balances.

(b) Include bank overdrafts and short-term loans, trade and other creditors, interest due, current taxation, short-term borrowing and any other short-term liabilities. Exclude deferred tax reserves.

(c) Include any long-term loans etc. raised in the UK and overseas other than from overseas members of your group of companies or your overseas head office.

(d) The net worth of the UK branch should equal the sum of questions 2.1, 2.2 & 2.3 minus questions 3.1, 3.2 & 3.3 and consists of its fixed assets (net of depreciation) plus its investments and its net current assets due to or from concerns other than the overseas head office or overseas members of its group of companies less long term loans other than those from the overseas head office or overseas members of its group of companies less deferred tax and deferred credits of the branch and less any provisions held by the branch. This should equal the unremitted profits and capital reserves of the branch, plus loans and net current account indebtedness due to the overseas head office or overseas members of its group of companies.

PLEASE KEEP THESE NOTES AND THE COPY OVERLEAF FOR REFERENCE

SC401C

A voluntary inquiry conducted by the Government Statistical Service

IN CONFIDENCE

DEPARTMENT OF TRADE AND INDUSTRY
Statistics Division 2
Room 252, Sanctuary Buildings
20 Great Smith Street
LONDON SW1P 3DB

Telephone: 01-215

Date 1 April 1987

	Ref	Sub	YE	UKInd	Tp	Roy	Co
1			86				

OT/IN/86S

← Please quote this reference number when writing to us

Please correct any errors in name or address

ANNUAL INQUIRY INTO TRANSACTIONS OF UNITED KINGDOM COMPANIES WITH THEIR OVERSEAS DIRECT INVESTORS IN 1986

Dear Contributor

I am writing to express my very warm appreciation of the contribution your company has made to this statistical inquiry in the past and to ask for your continued co-operation this year.

The purpose of this inquiry is to obtain up-to-date statistical information about various transactions between UK companies and their parent company overseas and overseas fellow subsidiaries and branches for use in compiling our country's balance of payments statistics. The information is needed by government in deciding its economic policy as well as its commercial policy in relation to other countries. This inquiry is the only source of such information and without it the accuracy of the UK balance of payments statistics will be greatly reduced.

The current form is overleaf and I would be most grateful for your co-operation in completing and returning it to me preferably not later than **30 June** this year.

All information supplied will be treated as confidential. Only those persons directly employed on collecting, processing and interpreting the information will have access to the particulars of individual returns, which will not be disclosed outside this small group of officials. The aggregated results which will be published next year in "Business Monitor, MA4, Overseas Transactions" will be compiled so as not to reveal the figures of individual contributors.

If you have any difficulties or would like any further information about this particular inquiry, my staff will be pleased to help you. The telephone extension of the person dealing with this inquiry is shown above.

Yours faithfully

P H Richardson.

P H RICHARDSON
Chief Statistician

SC106A

PLEASE RETURN THIS FORM BY 30 JUNE 1987

For official use

Ref		Sub	Form	
10			86	

		1	2
	161		

IMPORTANT Please refer to the enclosed notes before completing this form.
If you do not have precise figures readily available give the best estimates you can.
Please show all values in sterling rounded in £ million to one decimal place (e.g. £53,400,000 should be entered as 53.4). Use the same rates of conversion as in your own accounts.

1. DATE OF FINANCIAL YEAR END if not 31 December 1986 (see note 1) .... / /

2. COUNTRY IN WHICH OVERSEAS PARENT OR ASSOCIATE SHAREHOLDER IS INCORPORATED OR REGISTERED (see note 2) ....

3. OVERSEAS DIRECT INVESTMENT IN AND EARNINGS OF UK COMPANY

Item		Box	£ million	£ million
3.1 Overseas company's share of the UK reporting company's consolidated profit or loss (whether remitted or not) after allowing for provision for depreciation, tax and extraordinary items (see note 3.1)	a. Profit	162		
	or b. Loss	163		
3.2 Interest paid to, less interest received from, overseas parent company and overseas fellow subsidiaries and branches (see note 3.2)		164		
3.3 Dividends paid to overseas parent company (see note 3.3)		165		
3.4 UK tax credits paid directly by you to overseas parent company in respect of dividends (see note 3.4)		166		
3.5 Dealings by the overseas parent company in the reporting company's shares (including preference shares), debentures and quoted loan stock (see note 3.5)	a. Purchases	173		
	b. Sales	174		
3.6 Inter-company current account balances and loans outstanding between UK reporting company and the overseas parent company, overseas fellow subsidiaries, branches and associates (see note 3.6)	a. Net amount due from UK company at end of current financial year	176		
	or b. Net amount due to UK company at end of current financial year	177		
AND	c. Net amount due from UK company at end of previous financial year	178		
	or d. Net amount due to UK company at end of previous financial year	179		
3.7 Net gain or loss resulting from foreign currency exchange rate movements made by the UK reporting company on the balances reported under 3.6 (see note 3.7)	a. Gain	168		
	or b. Loss	169		

4. OTHER PAYMENTS AND RECEIPTS (see note 4)

Item		Box		
4.1 Royalties and similar transactions on revenue and capital account (see note 4.1)	a. Total amounts received less any tax paid overseas	203		
	b. Total amounts paid less any UK tax you deducted	204		
4.2 All other transactions not elsewhere specified (see note 4.2)	a. Total amounts received less any tax paid overseas	213		
	b. Total amounts paid less any UK tax you deducted	214		

Name of person to be contacted should this be necessary ------------------------
BLOCK CAPITALS PLEASE

Position in company ------------------------ Signature ------------------------

Tel No/Ext ------------------ Telex ------------------ Date ------------------

SC106E

NOTES ON COMPLETING FORM OT/IN/86/S

GENERAL

For the purpose of this return

a. the definition of a subsidiary is that given in Section 736 of the Companies Act 1985;

b. an overseas fellow subsidiary company or branch is a concern which has the same ultimate parent company as yourself but which possesses no direct shareholding in your company. If your immediate parent company is itself a subsidiary of another overseas company the ultimate overseas parent should be treated as a fellow subsidiary company.

c. a UK incorporated company is an associate of an overseas company if either the overseas company's interest in the UK company is effectively that of a partner in a joint venture or consortium, or the overseas company's direct interest in the UK company is long-term and substantial (i.e not less than 20% of the equity voting rights) and, having regard to the disposition of the other shareholdings, the overseas company is in a position to exercise a significant influence over the UK company

1. DATE OF FINANCIAL YEAR END

Where possible give details for 31 December 1986. If details are not readily available please give values for your annual accounting period ending after 30 June 1986 and before 1 July 1987, stating the date to which the figures relate.

2. COUNTRY IN WHICH OVERSEAS PARENT IS INCORPORATED OR REGISTERED

If you have more than one overseas investor please enter largest shareholder in first column and smaller investor in second column.

Include overseas fellow subsidiaries and branches with parent company
United Kingdom covers Great Britain and Northern Ireland.
Overseas is defined as being situated outside Great Britain, Northern Ireland, Isle of Man and the Channel Islands.

3. OVERSEAS DIRECT INVESTMENT IN AND EARNINGS OF UK COMPANY

3.1 The share of profits should be determined in accordance with the overseas company's proportionate holding of the reporting company's equity share capital. Contact us if you have any difficulties in determining this proportion (e.g. various classes of voting shares).

3.2 Interest paid should be net of any UK withholding tax you deducted. Interest received should be net of any tax paid overseas. If amounts received exceed that paid then enter figures as negatives.

3.3 Please give dividends paid in the year including dividends on preference shares.

3.4 If you have permission from the Board of Inland Revenue to pay tax credits direct to the overseas company under the UK double taxation agreements with Denmark, Luxembourg, the Netherlands, Finland, Switzerland, Sweden, Norway, Canada and the USA please show amounts paid in question 3.4.

3.5 Include transactions in quoted and unquoted shares and debentures of the reporting company plus loan stock listed on the UK Stock Exchange; purchases and sales of both existing and new issues of the reporting company's securities; acquisitions made against equipment, manufacturing rights, securities etc. supplied by the overseas company; acquisitions arising from the capitalisation of inter-company current account balances or unlisted loans (with a corresponding reduction in the reduction in the net amount due from you against question 3.6); purchases where an overseas company's shares are issued in exchange for your equity (value this at market value of the overseas company's shares on the day of purchase); and capital contributions from overseas company.

Exclude bonus issues of shares arising from the capitalisation of reserves.

3.6 State the net total of inter-company current account balances, long-term loans (other than debentures and quoted loan stock) and short-term loans outstanding between the UK reporting company and its overseas parent company and overseas fellow subsidiaries, branches and associates.

3.7 Include gain or loss made by the UK reporting company, resulting from foreign currency exchange rate movements, on inter-company current account balances, long-term and short-term loans reported in question 3.6 arising on amounts outstanding at the beginning or end of year.

4. ALL OTHER PAYMENTS AND RECEIPTS

It may not be possible to split payments or receipts between royalties and other transactions (e.g. where amounts include both royalties and other services such as advertising, market research, use of distribution and marketing facilities, etc.)
In this case enter the total amount in question 4.1 if royalties account for half or more, or in question 4.2 if royalties account for less than half.

Exclude
Payments for UK imports and exports, and their shipping and freight costs; and for cinematograph films and TV programmes

4.1 **Include**
Royalties on printed matter, sound recordings, performing rights etc., patent, licences, trademarks, designs, copyrights, mining and quarrying mineral rights, manufacturing rights, research and develpoment, technical know-how and assistance.
Any lump-sum settlements made in year.
Amounts payable to overseas agents for charges (less discounts) for services rendered in connection with royalties and similar transactions.

Exclude
Royalties, etc., included in the invoiced price of goods which you bought or sold.
Royalties, rentals and receipts for cinematograph films and TV programmes.

4.2 **Include**
Management and head office expenses, directors' fees, agents' commission on UK imports and exports; advertising, consultancy and design fees; legal services, trustee fees; gross profit or loss on goods bought and sold overseas without entering the UK; insurance premiums and claims, etc.

PLEASE KEEP THESE NOTES AND THE COPY OVERLEAF FOR REFERENCE

SC108C

PLEASE KEEP THIS COPY FOR REFERENCE

ANNUAL INQUIRY INTO TRANSACTIONS OF UNITED KINGDOM COMPANIES WITH THEIR OVERSEAS DIRECT INVESTORS IN 1986

IMPORTANT **Please refer to the enclosed notes before completing this form.**
If you do not have precise figures readily available give the best estimates you can.
Please show all values in sterling rounded in £ million to one decimal place (e.g. £53,400,000 should be entered as 53.4). Use the same rates of conversion as in your own accounts.

1. **DATE OF FINANCIAL YEAR END** if not 31 December 1986 (see note 1) .... / /

2. **COUNTRY IN WHICH OVERSEAS PARENT OR ASSOCIATE SHAREHOLDER IS INCORPORATED OR REGISTERED** (see note 2) ....

3. **OVERSEAS DIRECT INVESTMENT IN AND EARNINGS OF UK COMPANY**

Item		Box	£ million	£ million
3.1 **Overseas company's share of the UK reporting company's consolidated profit or loss** (whether remitted or not) after allowing for provision for depreciation, tax and extraordinary items (see note 3.1)	a. Profit	162		
	or b. Loss	163		
3.2 **Interest paid to, less interest received from, overseas parent company and overseas fellow subsidiaries and branches** (see note 3.2)		164		
3.3 **Dividends paid to overseas parent company** (see note 3.3)		165		
3.4 **UK tax credits paid directly by you** to overseas parent company in respect of dividends (see note 3.4)		166		
3.5 **Dealings by the overseas parent company in the reporting company's shares (including preference shares), debentures and quoted loan stock** (see note 3.5)	a. Purchases	173		
	b. Sales	174		
3.6 **Inter-company current account balances and loans** outstanding between UK reporting company and the overseas parent company, overseas fellow subsidiaries, branches and associates (see note 3.6)	a. Net amount due from UK company at end of current financial year	176		
	or b. Net amount due to UK company at end of current financial year	177		
AND	c. Net amount due from UK company at end of previous financial year	178		
	or d. Net amount due to UK company at end of previous financial year	179		
3.7 **Net gain or loss resulting from foreign currency exchange rate movements** made by the UK reporting company on the balances reported under 3.6 (see note 3.7)	a. Gain	168		
	or b. Loss	169		

4. **OTHER PAYMENTS AND RECEIPTS** (see note 4)

Item		Box	£ million	£ million
4.1 **Royalties and similar transactions** on revenue and capital account (see note 4.1)	a. Total amounts received less any tax paid overseas	203		
	b. Total amounts paid less any UK tax you deducted	204		
4.2 **All other transactions** not elsewhere specified (see note 4.2)	a. Total amounts received less any tax paid overseas	213		
	b. Total amounts paid less any UK tax you deducted	214		

Name of person to be contacted should this be necessary ------------------------
BLOCK CAPITALS PLEASE

Position in company ------------------------ Signature ------------------------

Tel No/Ext ---------------- Telex -------------- Date ----------------

SC106D

A voluntary inquiry conducted by the Government Statistical Service

IN CONFIDENCE

DEPARTMENT OF TRADE AND INDUSTRY
Statistics Division 2
Room 252, Sanctuary Buildings
20 Great Smith Street
LONDON SW1P 3DB

Telephone: 01-215

Date 1 April 1987

Ref		Sub	YE	UKInd	Tp	Roy	Co
1		86					

OT/IN/86B

← Please quote this reference number when writing to us

Please correct any errors in name or address

ANNUAL INQUIRY INTO TRANSACTIONS OF UNITED KINGDOM BRANCHES WITH THEIR OVERSEAS HEAD OFFICE IN 1986

Dear Contributor

I am writing to express my very warm appreciation of the contribution your company has made to this statistical inquiry in the past and to ask for your continued co-operation this year.

The purpose of this inquiry is to obtain up-to-date statistical information about various transactions between UK branches of overseas concerns and their overseas head office and overseas fellow subsidiaries and branches for use in compiling our country's balance of payments statistics. The information is needed by government in deciding its economic policy as well as its commercial policy in relation to other countries. This inquiry is the only source of such information and without it the accuracy of the UK balance of payments statistics will be greatly reduced.

The current form is overleaf and I would be most grateful for your co-operation in completing and returning it to me preferably not later than **30 June** this year.

All information supplied will be treated as confidential. Only those persons directly employed on collecting, processing and interpreting the information will have access to the particulars of individual returns, which will not be disclosed outside this small group of officials. The aggregated results which will be published next year in "Business Monitor, MA4, Overseas Transactions" will be compiled so as not to reveal the figures of individual contributors.

If you have any difficulties or would like any further information about this particular inquiry, my staff will be pleased to help you. The telephone extension of the person dealing with this inquiry is shown above.

Yours faithfully

P H Richardson.

P H RICHARDSON
Chief Statistician

SC105A

PLEASE RETURN THIS FORM BY 30 JUNE 1987

For official use

	Ref	Sub		Form
8			86	1

121 [ ]

IMPORTANT **Please refer to the enclosed notes before completing this form.**
If you do not have precise figures readily available give the best estimates you can.
Please show values in sterling rounded in £ million to one decimal place (e.g. £53,400,00 should be entered as 53.4). Use the same rates of conversion as in your own accounts.

1. DATE OF FINANCIAL YEAR END if not 31 December 1986 (see note 1) .... [/ /]

2. OVERSEAS COUNTRY IN WHICH HEAD OFFICE IS INCORPORATED OR REGISTERED (see note 2) .... [ ]

3. OVERSEAS DIRECT INVESTMENT AND EARNINGS.

£ million

3.1 **Total profit or loss of the branch** (whether remitted abroad or not) after allowing for provision for depreciation, UK tax and extraordinary items (see note 3.1).

a. Profit .... 122 [ ]
or
b. Loss .... 123 [ ]

3.2 **Total value of net worth of branch** attributable to overseas head office plus loans and current account balances of branch due to or from overseas fellow subsidiaries and branches (see note 3.2).

a. Net amount due from UK branch at end of current financial year .... 125 [ ]
or
b. Net amount due to UK branch at end of current financial year 126 [ ]

AND

c. Net amount due from UK branch at end of previous financial year .... 127 [ ]
or
d. Net amount due to UK branch at end of previous financial year .... 128 [ ]

3.3 **Net gain or loss resulting from foreign currency exchange rate movements** made by the branch on the net worth of branch reported in question 3.2 (see note 3.3).

a. Gain .... 171 [ ]
or
b. Loss .... 172 [ ]

4. OTHER PAYMENTS AND RECEIPTS (see note 4)

4.1 **Royalties and similar transactions** on revenue and capital account (see note 4.1).

a. Total amount received less any tax paid overseas .... 153 [ ]
b. Total amount paid less any UK tax you deducted .... 154 [ ]

4.2 **All other transactions not elsewhere specified** (see note 4.2).

a. Total amounts received less any tax paid overseas .... 157 [ ]
b. Total amounts paid less any UK tax you deducted .... 158 [ ]

Name of person to be contacted should this be necessary ____________________
BLOCK CAPITALS PLEASE

Position in company ____________________ Signature ____________________

Telephone No/Ext ____________________ Telex ____________ Date ____________

SC105B

PLEASE KEEP THIS COPY FOR REFERENCE

ANNUAL INQUIRY INTO TRANSACTIONS OF UNITED KINGDOM BRANCHES WITH THEIR OVERSEAS HEAD OFFICE IN 1986

IMPORTANT **Please refer to the enclosed notes before completing this form.**
If you do not have precise figures readily available give the best estimates you can.
Please show values in sterling rounded in £ million to one decimal place (e.g. £53,400,000 should be entered as 53.4). Use the same rates of conversion as in your own accounts.

1. **DATE OF FINANCIAL YEAR END if not 31 December 1986** (see note 1) .... / /

2. **OVERSEAS COUNTRY IN WHICH HEAD OFFICE IS INCORPORATED OR REGISTERED** (see note 2)........

3. **OVERSEAS DIRECT INVESTMENT AND EARNINGS.**

		Box	£ million
3.1 **Total profit or loss of the branch** (whether remitted abroad or not) after allowing for provision for depreciation, UK tax and extraordinary items (see note 3.1)	a. Profit	122	
	or		
	b. Loss........	123	
3.2 **Total value of net worth of branch** attributable to overseas head office plus loans and current account balances of branch due to or from overseas fellow subsidiaries and branches (see note 3.2).	a. Net amount due from UK branch at end of current financial year........	125	
	or		
	b Net amount due to UK branch at end of current financial year	126	
AND			
	c. Net amount due from UK branch at end of previous financial year.....	127	
	or		
	d. Net amount due to UK branch at end of previous financial year........	128	
3.3 **Net gain or loss resulting from foreign currency exchange rate movements** made by the branch on the net worth of branch reported in question 3.2 (see note 3.3)	a. Gain........	171	
	or		
	b. Loss........	172	

4. **OTHER PAYMENTS AND RECEIPTS** (see note 4)

		Box	
4.1 **Royalties and similar transactions** on revenue and capital account (see note 4.1).	a. Total amount received less any tax paid overseas........	153	
	b. Total amount paid less any UK tax you deducted........	154	
4.2 **All other transactions not elsewhere specified** (see note 4.2).	a. Total amounts received less any tax paid overseas........	157	
	b. Total amounts paid less any UK tax you deducted........	158	

Name of person to be contacted should this be necessary - - - - - - - - - - - - - - - -
BLOCK CAPITALS PLEASE

Position in company - Signature - - - - - - - - - - - - - - - - - - -

Telephone No/Ext - Telex - - - - - - - - - - - - - - Date - - - - - - - - - - - -

SC105D

NOTES ON COMPLETING FORM OT/IN/86B

GENERAL

For the purpose of this return:

(a) a branch means a permanent establishment as defined for the purpose of UK tax and UK double taxation agreements, being a fixed place of business in the UK through which the business of the overseas enterprise is wholly or partly carried on, and which is not a UK company (another questionnaire OT/IN/S covers UK companies in which overseas companies hold a significant shareholding),

(b) an overseas fellow subsidiary company or branch is a concern overseas which has the same ultimate parent company as yourself. If your overseas head office is itself a subsidiary of another overseas company the ultimate overseas parent should be treated as a fellow subsidiary company;

1. DATE OF FINANCIAL YEAR END

Where possible give details for 31 December 1986 If details are not readily available please give values for your annual accounting period ending after 30 June 1986 and before 1 July 1987, stating the date to which the figures relate.

2. OVERSEAS COUNTRY IN WHICH HEAD OFFICE IS INCORPORATED OR REGISTERED

United Kingdom covers Great Britain and Northern Ireland
Overseas is defined as being situated outside Great Britain, Northern Ireland, Isle of Man and the Channel Islands

3. OVERSEAS DIRECT INVESTMENT AND EARNINGS

3.1 Include in profit any interest payments to overseas head office and to overseas fellow subsidiaries and branches.

3.2 The total value of the net worth of the UK branch consists of the unremitted profits and capital reserves of the branch plus loans and net current account indebtedness due to the overseas head office and overseas fellow subsidiaries and branches. This should equal its fixed assets (net of depreciation) and its investments, plus its net current assets with concerns other than the overseas head office or overseas fellow subsidiaries and branches, less long term loans other than those from the overseas head office or overseas fellow subsidiaries and branches, less deferred tax and deferred credits of the branch, and less any provisions held by the branch

3.3 State the net gain or loss made by the UK branch, resulting from foreign currency exchange rate movements, on the net worth of the branch reported in question 3.2 arising on amounts outstanding at beginning or end of year

4. OTHER PAYMENTS AND RECEIPTS

It may not be possible to split a payment or receipt between royalties and other transactions (e.g. where amounts include both royalties and other services such as advertising, market research, use of distribution and marketing facilities, etc.). In this case enter the total amount in question 4.1 if royalties account for half or more, or in question 4.2 if royalties account for less than half

Exclude

Payments for UK imports and exports, and their shipping and freight costs, and for cinematograph films and TV programmes.

4.1 **Include**

Royalties on printed matter, sound recordings, performing rights etc., patents, licences, trademarks, designs, copyrights, mining and quarrying mineral rights, manufacturing rights, research and development, technical know-how and assistance
Any lump-sum settlements made in year.
Amounts payable to overseas agents for charges (less discounts) for services rendered in connection with royalties and similar transactions.
Total value of royalties etc. which you collect from, or pay to, overseas residents even though you may be acting as an agent.

Exclude

Royalties, etc., included in the invoiced price of goods which you bought or sold.
Royalties, rentals, payments and receipts for cinematograph films and TV programmes.

4.2 **Include**

Management and head office expenses, directors' fees, agents' commission on UK imports and exports, advertising, consultancy and design fees; legal services, trustee fees; gross profit or loss on goods bought and sold overseas without entering the UK; insurance premiums and claims etc.
Interest received from head office and overseas fellow subsidiaries and branches

PLEASE KEEP THESE NOTES AND THE COPY OVERLEAF FOR REFERENCE

SC105C

A voluntary inquiry conducted by the Government Statistical Service

IN CONFIDENCE

DEPARTMENT OF TRADE AND INDUSTRY
Statistics Division 2
Room 252, Sanctuary Buildings
20 Great Smith Street
LONDON SWIP 3DB

Telephone. 01-215-3529/3530

Date: 1 April 1985

CA/IN/84S

Please correct any errors in name or address

ASSETS AND LIABILITIES AT END 1984
OF UK SUBSIDIARIES AND ASSOCIATES OF OVERSEAS COMPANIES

Dear Contributor

We are conducting this three-yearly inquiry to obtain information on the value of overseas companies' investment in their subsidiaries and associates in the United Kingdom. It is the only source of such data and is used in compiling the UK balance of payments. There is also considerable demand outside government for comprehensive information on the value of overseas investment.

A summary of the results will be published in British business. More detailed results will be published later in the Business Monitor, MA4, Overseas Transactions, 1984 Supplement.

All information you supply will be treated as confidential. It will not be disclosed outside this office and will be used only for compiling totals which will not disclose the figures for individual contributors.

If the total value of overseas investment in your company (as defined overleaf) is between minus £50,000 and plus £50,000, just tick the box indicated and return the form to us. You do not have to answer any other questions.

This form should be returned even where there is little or nothing to report against some of the questions. I should like to emphasise that if you do not have precise figures readily available we are happy to accept your best estimates.

I would be grateful if you could return the completed form to us by 31 July 1985, or as soon as possible afterwards.

If you have any difficulties my staff will be pleased to help you. The telephone extension of the person dealing with this inquiry is shown above.

Yours faithfully

P H Richardson.

P H RICHARDSON
Chief Statistician

PLEASE RETURN THIS FORM BY 31 JULY 1985

For official use		Ref	Sub	Type	YE	Ind	Size	A	B	C	Amend
* 1984	15										

601		
605		

IMPORTANT Please refer to the enclosed notes before completing this form.
If you do not have precise figures readily available give the best estimates you can.
Please show all values to the nearest £ thousand (e.g. £53,400 should be entered as 53).

If the total value of overseas investment in your company (questions 4.2a + 4.5a + 5.2−5.1) is between minus £50,000 and plus £50,000 tick this box ☐, then sign and return the form. You do not have to answer any other questions.

1. DATE OF RETURN if not 31 December 1984 (see note 1) ……

2. GENERAL PARTICULARS (see note 2)
If you have more than one overseas investor please enter the largest shareholder in the first column and the smaller investor in the second column.

2.1 Name of immediate parent or associate company ……			
2.2 Country in which immediate parent or associate company is incorporated ……			
2.3 Country in which ultimate parent company is incorporated (see note 2f) ……			
2.4 Type of business of UK reporting company ……			
2.5 Percentage of issued share capital of UK reporting company owned by the immediate overseas parent or associate company ……	602	%	%

EITHER complete questions 3 to 5 if you are a subsidiary of an overseas company
OR complete question 6 if you are an associate of an overseas company (see note 2)

UK SUBSIDIARY COMPANY
Consolidated balance sheet at book value of UK company including all subsidiary companies whether incorporated in the UK or overseas

		£ thousands	£ thousands
3. ASSETS			
3.1 Fixed assets (see note 3a) ……	611		
3.2 Current assets (see note 3b) ……	613		
3.3 Total assets (3.1 plus 3.2) ……	606		
4. LIABILITIES			
4.1 Current liabilities (see note 4a) ……	614		
4.2 Long-term liabilities (see note 4b)			
(a) owing to the immediate and ultimate parent companies and overseas fellow subsidiaries (see note 4c) ……	615		
(b) other (see note 4d) ……	616		
4.3 Deferred tax and any long-term provisions or reserves (see note 4e) ……	617		
4.4 Minority shareholders' interests in any of the UK company's subsidiaries consolidated in the above figures ……	610		
4.5 Value of share capital and reserves of UK company (including preference shares)			
(a) directly attributable to immediate overseas parent company (see note 4f) ……	635		
(b) attributable to other shareholders ……	620		
4.6 Total liabilities (sum of 4.1 to 4.5) Should equal total assets at 3.3 ……	607		
5. ADDITIONAL ITEMS Inter-company account balance with immediate and ultimate overseas parents and overseas fellow subsidiaries included in:			
5.1 current assets at 3.2 above ……	618		
5.2 current liabilities at 4.1 above ……	619		
6. UK ASSOCIATE COMPANY Value of reporting company's share capital (including preference shares) and reserves directly owned by immediate overseas company, plus loans and net current account balances due to overseas company ……	627		

Name of person to be contacted should this be necessary ------------------------
BLOCK CAPITALS PLEASE

Position in company ------------------------ Signature ----------------

Telephone No/Ext -------------------- Telex -------------- Date ----------

PLEASE KEEP THIS COPY FOR REFERENCE **CA/IN/84S**

Ref

ASSETS AND LIABILITIES AT END 1984
OF UK SUBSIDIARIES AND ASSOCIATES OF OVERSEAS COMPANIES

IMPORTANT Please refer to the enclosed notes before completing this form.
If you do not have precise figures readily available give the best estimates you can.
Please show all values to the nearest £ thousand (e.g. £53,400 should be entered as 53).

If the total value of overseas investment in your company (questions 4.2a + 4.5a + 5.2–5.1) is between minus £50,000 and plus £50,000 tick this box ☐, then sign and return the form. You do not have to answer any other questions.

1. DATE OF FINANCIAL YEAR END if not 31 December 1984 (see note 1) ....

2. GENERAL PARTICULARS (see note 2)
If you have more than one overseas investor please enter the largest shareholder in the first column and the smaller investor in the second column.

2.1 Name of immediate parent or associate company			
2.2 Country in which immediate parent or associate company is incorporated			
2.3 Country in which ultimate parent company is incorporated (see note 2f)			
2.4 Type of business of UK reporting company			
2.5 Percentage of issued share capital of UK reporting company owned by the immediate overseas parent or associate company	602	%	%

EITHER complete questions 3 to 5 if you are a subsidiary of an overseas company
OR complete question 6 if you are an associate of an overseas company (see note 2)

UK SUBSIDIARY COMPANY
Consolidated balance sheet at book value of UK company including all subsidiary companies whether incorporated in the UK or overseas

		£ thousands	£ thousands
3. ASSETS			
3.1 Fixed assets (see note 3a)	611		
3.2 Current assets (see note 3b)	613		
3.3 Total assets (3.1 plus 3.2)	606		
4. LIABILITIES			
4.1 Current liabilities (see note 4a)	614		
4.2 Long-term liabilities (see note 4b)			
(a) owing to the immediate and ultimate parent companies and overseas fellow subsidiaries (see note 4c)	615		
(b) other (see note 4d)	616		
4.3 Deferred tax and any long-term provisions or reserves (see note 4e)	617		
4.4 Minority shareholders' interests in any of the UK company's subsidiaries consolidated in the above figures	610		
4.5 Value of share capital and reserves of UK company (including preference shares)			
(a) directly attributable to immediate overseas parent company (see note 4f)	635		
(b) attributable to other shareholders	620		
4.6 Total liabilities (sum of 4.1 to 4.5) Should equal total assets at 3.3	607		
5. ADDITIONAL ITEMS Inter-company account balance with immediate and ultimate overseas parents and overseas fellow subsidiaries included in:			
5.1 current assets at 3.2 above	618		
5.2 current liabilities at 4.1 above	619		
6. UK ASSOCIATE COMPANY Value of reporting company's share capital (including preference shares) and reserves directly owned by immediate overseas company, plus loans and net current account balances due to overseas company	627		

Name of person to be contacted should this be necessary ------------------------------
BLOCK CAPITALS PLEASE

Position in company ------------------------------ Signature ----------------------

Telephone No/Ext ---------------------- Telex -------------- Date -----------

PLEASE KEEP THIS COPY FOR REFERENCE

NOTES ON COMPLETING FORM CA/IN/84S

1. DATE OF FINANCIAL YEAR END

Where possible give details for 31 December 1984. If details are not readily available please give values for your accounting period immediately ending after 30 June 1984 but before 1 July 1985 stating the date to which the figures relate.

2. GENERAL PARTICULARS

For the purpose of this return:

(a) a United Kingdom company is one incorporated in Great Britain, Northern Ireland, Isle of Man or the Channel Islands.

(b) your immediate parent or associate company is the concern which actually owns your share capital;

(c) the definition of a subsidiary company is that given in Section 154 of the Companies Act 1948;

(d) a UK incorporate company is an associate of an overseas company if either the overseas company's interest in the UK is effectively that of a partner in a joint venture or consortium, or the overseas company's direct interest in the UK company is long-term and substantial (i.e not less than 20% of the equity voting rights) and, having regard to the disposition of the other shareholdings, the overseas company is in a position to exercise a significant influence over the UK company;

(e) an overseas fellow subsidiary company is a concern overseas which has the same ultimate parent company as yourself but which possesses no direct shareholding in your company;

(f) If your immediate parent or associate company is a subsidiary of another company please state the country in which the ultimate parent company is incorporated.
If your immediate parent or associate company is not a subsidiary of another company write 'NONE' in the box.

3. ASSETS

(a) Include all tangible and intangible fixed assets (including goodwill) and investments in associates held by the UK reporting company after allowing for depreciation plus investment in securities. Deduct any government grants on these assets held as deferred credits or reserves.

(b) Include stocks, debtors, short-term loans, bank balances, cash and long-term loans by the UK company to immediate and ultimate overseas parent companies and to overseas fellow subsidiary companies.

4. LIABILITIES

(a) Include bank overdrafts and short-term loans, trade and other creditors, dividends and interest due, current taxation, short-term borrowing and other provisions of a current character. Exclude deferred taxation.

(b) Include any debentures, long-term loans, or other loans not included in current liabilities at 4.1.

(c) Include any long-term loans, etc, due to your immediate and ultimate parent companies and overseas fellow subsidiary companies including those in different overseas countries from your immediate parent company.

(d) Include any long-term loans, etc, raised in the UK and overseas other than those from your immediate and ultimate parent companies or from overseas fellow subsidiaries in the group.

(e) Exclude any reserves which are part of the shareholders' interest.

(f) Include share capital and reserves attributable to the immediate parent company and any held directly by the ultimate parent company and fellow subsidiaries. However, if you also have an associated investing company with the same ultimate parent as yourself for which you have given separate figures in the second column, you should enter its shareholding in the first column at 4.5(b) and also include it in the second column at question 6.

PLEASE KEEP THESE NOTES AND THE COPY OVERLEAF FOR REFERENCE

A **compulsory** inquiry conducted by the Government Statistical Service
IN CONFIDENCE

DEPARTMENT OF TRADE AND INDUSTRY
Statistics Division 2
Room 252, Sanctuary Buildings
20 Great Smith Street
LONDON SWIP 3DB

Telephone 01-215-3529/3530

Date 1 April 1985

CA/IN/84B

Please correct any errors in name or address

NOTICE UNDER SECTION 1 OF THE STATISTICS OF TRADE ACT 1947

The Secretary of State for Trade and Industry hereby requires you by law to provide the information called for in this inquiry form with the exception of questions 4.3 & 4.4 The information in this form is required for the purposes of the appreciation of economic trends and the provision of a statistical service for government and industry.

Please complete this form and return it to the Department of Trade and Industry by 2 September 1985.

NB: The information given by you under the Act will be treated as confidential in strict accordance with the Act and subject to the further restriction that information about individual businesses will be used and disclosed under Ministerial direction to other government departments for statistical purposes only, except that the names and addresses of individual businesses and their industrial classification may if a Minister so directs be made available to Ministers and officials of government departments who need them for carrying out their functions

ASSETS AND LIABILITIES AT END 1984
OF UK BRANCHES OF OVERSEAS COMPANIES

Dear Contributor

We are conducting this three-yearly inquiry to obtain information on the value of overseas companies' investment in their branches in the United Kingdom. It is the only source of such data and is used in compiling the UK balance of payments.

A summary of the results will be published in British business. More detailed results will be published later in the Business Monitor, MA4, Overseas Transactions, 1984 Supplement.

Please note that questions 4.3 & 4.4 are voluntary and you are not required by law to answer them. However it would be very helpful if you could give this information as it will complete the balance sheet for the branch. Any information given voluntarily will be treated as confidential. It will not be disclosed outside this office and will be used only for compiling totals which will not disclose the figures for individual contributors.

If the total value of overseas investment in your branch (as defined overleaf) is between minus £10,000 and plus £10,000 just tick the box in question 2.1 and return the form to us. You do not have to answer any other questions.

This form should be returned even where there is little or nothing to report against some of the questions. I should like to emphasise that if you do not have precise figures readily available we are happy to accept your best estimates.

I would be grateful if you could return the completed form to us as soon as possible. In any case it must be returned by 2 September 1985.

If you have any difficulties my staff will be pleased to help you. The telephone extension of the person dealing with this inquiry is shown above.

Yours faithfully

P H Richardson

P H RICHARDSON
Chief Statistician

PLEASE RETURN THIS FORM BY 2 SEPTEMBER 1985

For official use			Ref	Sub	Type	YE	Ind	Size	A	B	C	Amend
*	1984	16										

601	

IMPORTANT Please refer to the enclosed notes before completing this form.
If you do not have precise figures readily available give the best estimates you can.
Please show all values to the nearest £ thousand (e.g. £53,400 should be entered as 53).

1. DATE OF RETURN if not 31 December 1984 (see note 1) ....

2. GENERAL PARTICULARS (see note 2)

2.1 If your net assets (question 3 minus questions 4.1 & 4.2) are between minus £10,000 and plus £10,000, tick this box ☐ then sign and return the form. You do not have to answer any other questions.

2.2 Country in which overseas company is incorporated ....

2.3 Trading name of overseas company if different from trading name in UK..........

2.4 Type of business of UK branch ....

2.5 If you are resident in the UK for tax purposes and have UK subsidiary companies, please list them here.

....

....

AND EITHER include the value of your investment in the UK subsidiary companies in question 3.2 and tick this box .... ☐ | 631 | 2 |

OR if more convenient, give combined figures for yourself and your UK subsidiaries in each part of questions 3 & 4 and tick this box .... ☐ | 631 | 1 |

If you do not own all the share capital of a subsidiary company include the minority shareholders' interests in "other liabilities" in question 4.3.

3. ASSETS

		£ thousands
3.1 Fixed assets after provision for depreciation (see note 3a)..........	621	
3.2 Investments in company and government securities and loans(see note 3b)	622	
3.3 Current assets (see notes 3b & 3c)	623	

4. LIABILITIES

4.1 Current liabilities (see notes 4a & 4b)	624	
4.2 Long-term liabilities (see notes 4a & 4c)..........	625	
4.3 Provisions, deferred tax reserves and other liabilities (voluntary question)..	626	
4.4 Net worth of UK branch (see note 5) (voluntary question)..........	636	

Name of person to be contacted should this be necessary ------------------------------
BLOCK CAPITALS PLEASE

Position in company -------------------------- Signature ------------------

Telephone No/Ext ---------------------Telex -------------- Date ----------

PLEASE KEEP THIS COPY FOR REFERENCE **CA/IN/84B**

Ref

ASSETS AND LIABILITIES AT END 1984
OF UK BRANCHES OF OVERSEAS COMPANIES

IMPORTANT Please refer to the enclosed notes before completing this form.
If you do not have precise figures readily available give the best estimates you can.
Please show all values to the nearest £ thousand (e.g. £53,400 should be entered as 53).

1. DATE OF RETURN if not 31 December 1984 (see note 1) ....

2. GENERAL PARTICULARS (see note 2)

2.1 If your net assets (question 3 minus questions 4.1 & 4.2) are between minus £10,000 and plus £10,000, tick this box ☐ then sign and return the form. You do not have to answer any other questions.

2.2 Country in which overseas company is incorporated..........

2.3 Trading name of overseas company if different from trading name in UK....

2.4 Type of business of UK branch ....

2.5 If you are resident in the UK for tax purposes and have UK subsidiary companies, please list them here.

....

....

AND EITHER include the value of your investment in the UK subsidiary companies in question 3.2 and tick this box .... ☐ | 631 | 2 |

OR if more convenient, give combined figures for yourself and your UK subsidiaries in each part of questions 3 & 4 and tick this box .... ☐ | 631 | 1 |

If you do not own all the share capital of a subsidiary company include the minority shareholders' interests in "other liabilities" in question 4.3.

3. ASSETS

		£ thousands
3.1 Fixed assets after provision for depreciation (see note 3a)	621	
3.2 Investments in company and government securities and loans (see note 3b)	622	
3.3 Current assets (see notes 3b & 3c)	623	

4. LIABILITIES

4.1 Current liabilities (see notes 4a & 4b)	624	
4.2 Long-term liabilities (see notes 4a & 4c)	625	
4.3 Provisions, deferred tax reserves and other liabilities (voluntary question)	626	
4.4 Net worth of UK branch (see note 5) (voluntary question)	636	

Name of person to be contacted should this be necessary ----------
BLOCK CAPITALS PLEASE

Position in company ---------- Signature ----------

Telephone No/Ext ---------- Telex ---------- Date ----------

PLEASE KEEP THIS COPY FOR REFERENCE

NOTES ON COMPLETING FORM CA/IN/84B

1. DATE OF RETURN

Where possible give details for 31 December 1984. If details are not readily available please give values for your annual accounting period ending after 30 June 1984 and before 1 July 1985, stating the date to which the figures relate.

2. GENERAL PARTICULARS

(a) If you are a company incorporated in Great Britain, Northern Ireland, Isle of Man or the Channel Islands please state this against question 2.2 and return the form to us uncompleted.

For the purpose of this return

(b) A UK branch means a permanent establishment as defined for the purposes of UK tax and UK double taxation agreements, being a place of business in the UK through which the business of the overseas enterprise is wholly or partly carried on and which is not a company incorporated in Great Britain, Northern Ireland, Isle of Man or the Channel Islands.

(c) Overseas members of your group of companies means companies which are:

i. directly or indirectly owned by your company, plus, where the company of which you are a branch is itself a subsidiary, your ultimate parent company and all its directly and indirectly owned subsidiaries; and

ii. incorporated outside Great Britain, Northern Ireland, Isle of Man and the Channel Islands.

(d) For overseas incorporated companies with their head office in the UK, 'overseas head office' means the registered office of the company in its country of incorporation.

3. ASSETS

(a) Include all tangible and intangible fixed assets in the UK owned by the company which are used or employed for the business of the branch. Deduct the value of any government grants on fixed assets held as deferred credits.

(b) Exclude from questions 3.2 & 3.3 any loans to and other amounts due from overseas members of your group of companies, and overseas head office and inter-branch balances.

(c) Include stocks, debtors, short-term loans, bank balances, prepayments (deferred assets) and cash.

4. LIABILITIES

(a) Exclude from questions 4.1 and 4.2, loans from and other amounts due to overseas members of your group of companies, overseas head office and inter-branch balances.

(b) Include bank overdrafts and short-term loans, trade and other creditors, interest due, current taxation, short-term borrowing and any other short-term liabilities. Exclude deferred tax reserves.

(c) Include any long-term loans etc. raised in the UK and overseas other than from overseas members of your group of companies or your overseas head office.

5. VOLUNTARY QUESTIONS

The net worth of the UK branch should equal question 3 less questions 4.1, 4.2 & 4.3 and consists of its fixed assets (net of depreciation) plus its investments and its net current assets due to or from concerns other than the overseas head office or overseas members of its group of companies less long term loans other than those from the overseas head office or overseas members of its group of companies less deferred tax and deferred credits of the branch and less any provisions held by the branch. This should equal the unremitted profits and capital reserves of the branch, plus loans and net current account indebtedness due to the overseas head office or overseas members of its group of companies.

PLEASE KEEP THESE NOTES AND THE COPY OVERLEAF FOR REFERENCE

SUBJECT INDEX